HOW
TO USE
METERS

by John F. Rider
and Sol D. Prensky
Assistant Professor in Electrical Engineering
Fairleigh Dickinson University

Second Edition

Revised

John F. Rider Publisher, Inc., New York
Chapman & Hall, Ltd., London

PREFACE

This revision of the second edition of *How to Use Meters* is twofold: (1) the original text has been brought up to date, and (2) new material has been added on modern advances in the rapidly growing field of meter use. Every effort has been made to retain the original point of view which proved so successful in the first edition, i.e., satisfying the practical interest of those concerned with measuring devices and their basic principles. As the electronic art expands in breadth and depth of applications, it becomes increasingly important to preserve this most effective approach by placing the accent on fundamental meter function.

This approach has been kept in mind in introducing such new developments as transistorized voltmeters, refined laboratory instruments providing greatly increased sensitivities, and advanced indication features in meters providing long-arc (250°) meter scales and digital displays. Careful attention has been given to the important extension of measurement capability in the new chapters on specialized measurements (Chap. 14) and advanced meter features (Chap. 15). Circuits are discussed to illustrate the use of electrometer-type voltmeters having extremely high input impedances, and of chopper-type stabilized galvanometers capable of measuring minute d-c currents. Summary tables have been added to allow at-a-glance comparisons of the numerous meter types. Attention has also been paid to the fast-growing areas in industrial electronics where transducers are widely used to display and control physical quantities after they have been converted into electrical values.

It is recognized that the inclusion of too many specialized details and complexities could easily tend to obscure the picture of the fundamental functioning of the measuring devices with which this text is primarily concerned. To avoid this tendency, a reference reading list of specialized texts is given in Chap. 14, to direct further study in the more special fields of electronic instruments.

Grateful acknowledgment is made, as before, to the many manufacturers who supplied information and illustrations, as indicated by the credit lines that accompany their very helpful material.

New York, N. Y.
December, 1959

JOHN F. RIDER
SOL D. PRENSKY

CONTENTS

TABLES

1

PRINCIPLE AND CONSTRUCTION OF D-C MOVING COIL METERS

1-1. Introduction

Electric meters are devices which *measure* electric current and other electrical quantities and *indicate* the quantity measured. The heart of an electric meter is the *meter movement.* The movement translates electrical energy into mechanical energy which causes visual indication. The indicator usually is a pointer moving across a calibrated scale.

A number of different types of meter movements have been designed and are in use. The most important of these is the *moving-coil* movement, which in its original form was developed by the French scientist *D'Arsonval.* The modern moving-coil movement is the result of the work of *Dr. Edward Weston,* who considerably improved on *D'Arsonval's* original version. In its present-day form moving-coil movement is referred to as either a *Weston-type* or a *D'Arsonval* movement. A majority of the meters discussed in this book use this type of mechanism.

1-2. Principles of the Moving-Coil Meter Movement

This type of meter depends for its operation on the reaction between a direct current-carrying coil and a stationary magnetic field. The magnetic lines of force generated by the current in the wire interact with the unvarying lines of force of the magnetic field in which the coil is located, and the coil is forced to *move* with respect to the stationary field. By virtue of its physical design, the coil rotates. In order that there be useful interaction, the current in the coil must be unidirectional, hence the instrument is a d-c meter.

Suppose a *loop* of wire is located in a magnetic field (Fig. 1-1A) in which the poles of the magnet are concave, so that the loop of wire

1

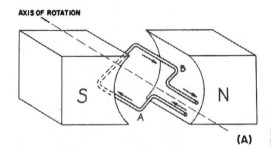

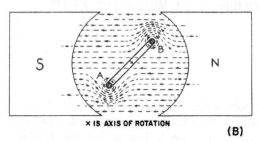

Fig. 1-1. Illustration of the basic motor principle employed in moving-coil permanent-magnet meters.

can revolve freely. Direct current is made to flow through the coil. The result is shown at (B) of Fig. 1-1, where the cross-sectional view indicates the magnetic lines of force. Note that the lines of force from the magnet and wire are in the same direction *above*[1] wire B, so wire B is forced *downward*. They also run in the same direction *below* wire A, so wire A is forced *upward*. By making the coil of many turns the effect is increased, and the same rotational force (torque) can be obtained with a smaller current in the coil.

1-3. D'Arsonval Moving-Coil Mechanism

The construction of the moving-coil meter mechanism is shown in Fig. 1-2. The permanent magnet which provides the stationary field is equipped with extensions, called "pole pieces," shaped to fit around the moving-coil assembly. The moving coil is wound on a thin, light, non-magnetic metal form attached to pivots which rotate on jeweled bearings. Helical springs are attached to the ends of the coil and are also used to carry current from the external source to the moving coil. An iron core is mounted inside the coil so that the magnetic path from the main horseshoe, permanent magnet is entirely through iron, except for the annular gap in which the thin sides of

---

[1] The direction of lines of force around the wire can be determined from the "left-hand rule" which says that if the wire is grasped in the left hand with the thumb extended along the wire in the direction of electron flow, the fingers will point around the wire in the direction of the magnetic lines of force.

the moving coil rotate. The pointer is mounted on the moving-coil assembly so that it moves with the moving coil. Mechanical stops are mounted on the meter to limit the movement of the pointer to its desired range.

When a current is passed through the coil, interaction between the magnetic fields of the coil and magnet rotates the moving-coil assembly clockwise against the resistance of the springs. When the force of resistance from the springs becomes equal to the force of rotation from the action of the magnetic fields, the moving coil and the attached pointer stop. How far the pointer will move to the right depends upon the amount of current in the coil. By using a *scale* across which the upper end of the pointer can move, we can read how

much current is flowing through the coil. The indicating pointer is attached to the coil assembly and swings with it, as is shown in the assembly construction view of Fig. 1-3. There is a screw on the front of the movement to permit adjusting the pointer to its mechanical zero for zero current (see Fig. 1-2D).

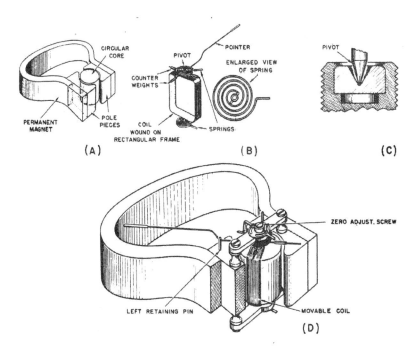

Fig. 1-2. Construction details of a D'Arsonval moving-coil mechanism.

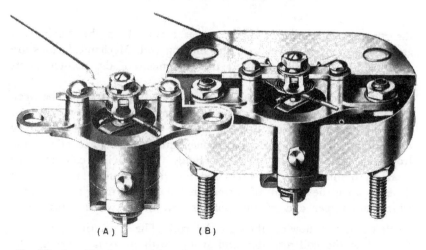

Fig. 1-3. Assembly of commercial moving-coil meter: (A) full double-bridge suspension of D'Arsonval movement; (B) movement assembled on permanent magnet without scale.
Pace Elec. Inst. Div., Precision Apparatus Co.

1-4. Coil Resistance

We noted in Sec. 1-2 that an increase in the number of turns in the coil reduces the amount of current necessary to produce a given turning force, and thus deflection, of the pointer. Accordingly, meters designed for the lowest full-scale deflection current must have the greatest number of turns of wire on the coil, and are referred to as the "most sensitive" meters. The lower the full-scale current rating of a meter, the greater is its sensitivity. But whatever the sensitivity rating, each moving-coil meter also bears its own d-c resistance rating.

Since sensitive meters require more turns of wire they also have the greatest coil resistance. The resistance of a meter is a very important thing to know for, as will be explained later, the resistance of a meter determines its usefulness in making practical current and voltage measurements in electronic circuits.

1-5. Scale Calibration for Moving-Coil Type Meter

The force tending to rotate the moving coil assembly is proportional to the coil current. At the same time, the resistance torque of the springs is directly proportional to the degrees of rotation from the zero position. These two facts lead to the conclusion that the displacement of the pointer along the meter scale is directly proportional to the current. When a relation of this type exists, we say that the scale is *linear*. A linear scale means that the same *change* in current always produces the same change in pointer position no matter in what part of the scale the pointer is when the change is made.

Figure 1-4 shows a typical linear meter scale, 150 milliamperes d-c full scale. Notice that this scale is labeled every 30 ma. There are five divisions for every 10 ma; each division representing 2 ma each. Now note that the separation along the scale from zero to 10 ma is the same as the distance from 30 to 40 ma, from 60 to 70 ma, 80 to 90 ma, 140 to 150 ma and between any two points 10 ma apart.

The linear scale is characteristic of the d-c moving-coil permanent-magnet meter movement, as compared to other types, which employ non-linear scales. These non-linear types are shown in Chapter 2 and elsewhere in this book.

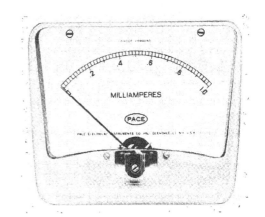

Fig. 1-4 The linear scale used with a typical moving-coil meter (0–1 ma). Pace Elec. Inst., model 45P, (clear acrylic case)

1-6. Importance of Polarity

The steady magnetic field in which the moving coil operates is supplied by the permanent magnet. This field is constant and always in one direction. In other words the North pole remains North and the South pole remains South. For this reason, the current in the coil must always have the proper polarity, or flow in the proper direction to make the pointer deflect normally (from left to right).

Therefore, for the instrument to give the desired indication, the polarity of the terminals of a moving coil (d-c) meter must be taken into account in making connections. On most instruments this polarity is indicated on the case, near each terminal or on the terminal itself. Frequently only the positive terminal is marked with a + sign; then, of course, the other terminal must be the negative one. The meter must be so connected that the electron flow is *into* the *negative* terminal and *out* of the *positive* terminal.

**Fig. 1-5. A Weston center zero
galvanometer.**

The moving-coil permanent-magnet (d-c) meter *cannot* be used
to measure alternating or rapidly-varying direct currents and voltages
without the use of auxiliary equipment, such as a rectifier. When alternating current is passed through the coil of the d-c meter, the coil's
magnetic field changes polarity with each new alternation of the
current. The coil assembly and the pointer thus tend to rotate back
and forth in synchronism with the current's alternations. Although at
power frequencies as low as 25 cps some vibration of the pointer may
be noticeable, at 60 cps and higher the inertia of the system prevents
the pointer from moving at all.

1-7. Center-Zero Galvanometer

The term "galvanometer" has taken on a special meaning. It is
used to distinguish a certain type of d-c meter having two main characteristics of the original D'Arsonval moving-coil instrument. These
are great sensitivity, and zero-current indication at the *center* rather
than at the left end of the scale.

In the galvanometer, the moving coil is suspended in such a way
that it can move the pointer in either direction from its center position. It is used primarily in applications in which it is desired that a
balance be indicated; in other words, applications in which normal
current is zero, and deviation from zero in either the positive or the
negative direction is to be observed. An example of this instrument
is shown in Fig. 1-5.

1-8. Long-arc (250°) Meter Scale

A further increase in scale readability is provided by the development of the 250°-arc d-c meter scale. This long-arc type of scale is
illustrated in Fig. 1-6. These meters, using a modified form of the

permanent-magnet moving-coil mechanism, provide a longer calibrated arc (generally 250°) for the pointer movement, as compared to the most commonly encountered 90°-arc type of scale. This results in a substantial increase in scale length for a meter of a given size and a corresponding increase in readability of the scale. Thus, for example, in the popular 3½-in. diameter d-c meter, the face of the conventional 90°-arc meter provides 2.36 in. of scale length, while the same size of 250°-arc meter provides a scale length of fully 5 in., or better than twice the conventional scale length. Another incidental advantage that may be obeserved in the illustration is the ease with which the 250° scale lends itself to convenient division into 100 equal parts, thus adding further to its readability.

The construction details of the 250°-arc as used in a switchboard permanent-magnet moving-coil d-c meter are seen in Figs. 1-7 and 1-8. By comparison with the 90°-arc meter, it is apparent that the salient

Fig. 1-6. Long-arc (250°) meter
        scale. Weston

characteristics of the long-arc type is the manner in which the permanent-magnet field is established and shaped. The conventional 90°-arc type, when viewed from the front, shows the permanent magnet as an inverted U, supplying a magnetic field extending horizontally from left to right between the pole pieces. By contrast, the long-arc type viewed from the front shows a magnet structure having an eccentric S-shaped magnetic gap, with the magnetic lines of force extending perpendicularly into the paper from front to back. As the coil moves through its 250° rotation within this gap, it encounters a varying gap-width, designed in spiral form, so that the coil deflection remains linear with the torque and causes rotation throughout the full 250°

Fig. 1-7. Construction of 250°-arc meter (switchboard type). *Westinghouse*

swing. This form of magnetic field construction overcomes the theoretical limitation of 180° rotation inherent in the conventional form of magnetic field.

It should be noted that the advantage of obtaining greater readability through the use of longer scales—either by a larger size of meter in conventional construction, or by a greater arc length in the same

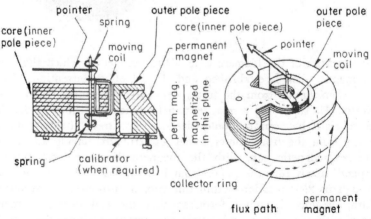

Fig. 1-8. Construction detail sketch of 250°-arc meter. *Westinghouse*

size meter in the long-arc construction—is achieved at an increased cost, as is to be expected. Disregarding the shape of the external meter case (whether round or square, shape is unimportant) the type of scale most commonly encountered for general-purpose service use continues to be the 90° arc of the $3\frac{1}{2}$-in. meter size, generally subdivided into 50 divisions and providing a scale length of slightly under $2\frac{1}{2}$ in. The long-arc type shown in Figs. 1-7 and 1-8 has long been in use on switchboard instruments, while the smaller long-arc type panel instrument (Fig. 1-6) is becoming increasingly popular wherever its extra cost is justified.

1-9. Galvanometers of Higher Sensitivity

Other types of galvanometers are available for obtaining a greater current sensitivity than that provided in the ordinary pivoted moving-coil movement, in which deflection from zero to either end of the scale usually requires from 25 to 500 μa. One such more-sensitive type is the string galvanometer which is often used in laboratories and provides a sensitivity high enough to produce a readable response to a current of a few μa. In this instrument, the moving coil is suspended between the poles of the permanent magnet by a delicate string, so that only a very small force is required to cause it to turn. A mirror mounted on the suspension string reflects light onto a scale placed at some distance (usually about a half a yard or meter) away from the mirror, so that the slight angular motion of the mirror is greatly magnified by the ·long path of the light reflected on to the scale. As a result, when properly adjusted, a very slight current of a few μa in the moving coil produces a readable optical deflection on the scale. However, because of the delicate adjustments necessary for the string suspension and the optical system, the high sensitivity of this type of galvanometer is limited only to those applications where the meter can be used in a stable, fixed position. Otherwise, where a combination of both high sensitivity and portability is required in a galvanometer, it is possible to make use of a d-c *electronic galvanometer*, which operates in an electrometer type of vacuum-tube circuit. By the liberal use of electronic amplifying and stabilizing circuits, the electronic galvanometer can be made to measure in the 10^{-9}-amp (millimicroampere or mμa) range, and under special conditions, even into the 10^{-12} (micromicroampere or $\mu\mu$a) range. This type is described in greater detail in Chap. 14.

1-10. Other Developments in D-C Meter Construction

Interesting developments in the construction of d-c meters include new "taut-band" suspension method, which results in greater sensitivity

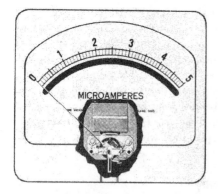

Fig. 1-9. Cutaway view of "taut-band" suspension-type meter. *Hickok*

for portable meters than is obtained by the more conventional spring suspension previously discussed. By completely eliminating the conventional pivots, bearing, and hairsprings, the taut-band suspension meter illustrated in Figs. 1-9 and 1-10 provides a more friction-free movement. As a result, sensitivities in the order of 2 μa full-scale are made available in portable meters that are comparable to the previously available 50-μa full-scale meters of conventional pivoted construction.

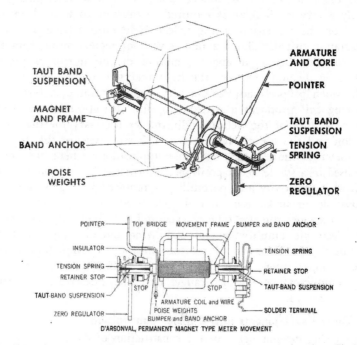

Fig. 1-10. Construction of "taut-band" suspension meter: (A) perspective view; (B) side view.

This increased sensitivity is obtained without sacrificing either overload capacity or reliability.

Another development that uses the magnet as the moving element rather than the coil is illustrated in Fig. 1-11. This moving-magnet mechanism is used for certain types of large-size panel instruments having full-scale sensitivities above 5 ma.

1-11. Classification of Moving-Coil Meters

The D'Arsonval meter, which is distinguished by its moving-coil permanent-magnet construction, has been singled out in this discussion of d-c meters as representative of meters for d-c use because of its

TABLE 1-1. A-C—D-C CLASSIFICATION OF COMMONLY USED METER TYPES

Meter type	Suitability	Major use
D'Arsonval (moving-coil)	d-c only	most widely-used meter for d-c current and voltage measurements in low- and medium-impedance circuits
Moving-iron	d-c or a-c	inexpensive type used for rough indication of large currents (like automobile-battery charging or corresponding voltage
Dynamometer	d-c or a-c	widely used for precise a-c voltage and current measurements at power frequencies (in low-impedance circuits)
Electrostatic	d-c or a-c	limited use in high-voltage measurements (at very high impedances)
Electronic (VTVM or transistor VM)	d-c or a-c	extensive laboratory use as voltmeter in circuits where both high-sensitivity and high-impedance requirements are combined
Rectifier* (combined with D'Arsonval movement)	a-c only	widely used for medium-sensitivity service-type voltage measurements in medium-impedance circuits

* Used in the volt-ohm-milliammeter (VOM), with switch for selecting d-c and a-c ranges.

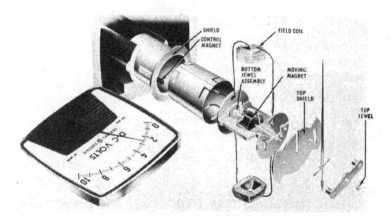

Fig. 1-11. Moving-magnet type of mechanism used in large-size panel instruments. GE type DO-91

predominance in d-c measurement. However, there are other types of meters that are able to be used on dc, and it will be profitable, at this point, to list various other outstanding types of meters (or instruments using these meters) and to distinguish between them on the basis of their suitability either for dc only, ac only, or for universal (a-c or d-c) use. Table 1-1 shows how the D-Arsonval type of meter fits in with the other types of instrument meters in common use.

2

PRINCIPLE AND CONSTRUCTION OF A-C METER MOVEMENTS

The measurement of an alternating current (or voltage) is the measurement of a continuously varying quantity. The d-c moving-coil meter movement can make this measurement if the frequency of the current is very low, up to a few cycles per second. At these low frequencies the moving coil can follow the changes in current value and direction instant by instant. However, if the frequency is higher, and most a-c meters are used at power frequencies which are between 50 and 60 cps, the moving-coil type of meter movement is not suitable for direct application and must be "adapted," as is described in Chapter 6. Hence a different type of meter movement is required.

2-1. Basic Principle of Moving-Iron Meters

If a soft iron bar is placed near a coil which is carrying current, the bar becomes magnetized, as illustrated in Fig. 2-1. The magnetic lines of force issuing from the coil pass through the bar in the same direction as the lines of force induced in the bar by the coil. Since lines of force behave like stretched rubber bands that try to shorten themselves as much as possible, the bar is attracted to the electromagnet. If the coil is fixed and the bar is free to move, the bar is pulled into the coil.

If the current in the coil reverses, the lines of force issuing from the coil reverse, and the lines of force in the bar also reverse. The lines of force in the bar line up with the lines of force coming from the coil, and the bar is again attracted to the coil. Therefore the soft iron bar is attracted to the electromagnet regardless of the direction of the current flowing in the coil, hence the magnetized bar is attracted to the coil with dc or ac flowing in the coil. Soft iron is used for the bar because it demagnetizes readily when the current through the coil ceases.

13

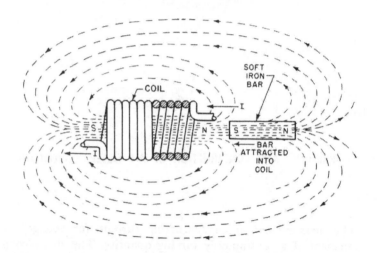

Fig. 2-1. Basic principle behind the action of the moving-iron type of meter.

When the coil illustrated in Fig. 2-1 carries the sine waveform current shown in Fig. 2-2, the direction of movement of the bar into the coil is the same during each half cycle or alternation. This is true in spite of the fact that the current reverses its direction of flow. The extent of the motion of the magnetized bar depends on how freely it can move and on the amount of current flowing in the coil, hence the intensity of the magnetizing force.

The fact that there is a repulsion between the induced field in a metal object and the alternating field inducing it is the basis of operation of a group of a-c meters. They are known as the *moving-iron* type of meter.

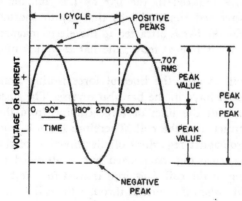

Fig. 2-2. Basic sine wave, showing how the current or/voltage builds up in one direction, returns to zero and then builds up in the other direction, finally returning to zero at the end of a complete cycle.

2-2. Moving-Iron Types of A-C Meters

These meters make use of the principle that a coil carrying current magnetizes a piece of soft iron which is placed within the coil's magnetic field. In practical devices two pieces of iron are used instead of one, and both are placed under the influence of the same magnetic field by being located within the coil. The principle is shown in Fig. 2-3.

When current flows through the coil the two bars are magnetized with the same polarity. Because these adjacent poles are of similar polarity, they repel each other. If the current reverses, as in Fig. 2-2, the polarity of both magnetized rods is reversed at the same time and they still repel each other. If one bar is fixed and the other is free to move, the force of repulsion can be made to indicate the amount of current flow by causing the movable bar to move a pointer across a calibrated scale.

Fig. 2-3. Illustration of the basic principle underlying the operation of the moving-vane meter. The electromagnet magnetizes both bars with the same polarity, so they repel each other with a force depending upon the strength of the current in the coil.

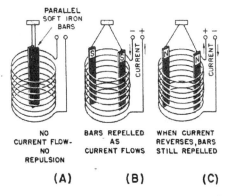

PARALLEL SOFT IRON BARS

(A) NO CURRENT FLOW- NO REPULSION

(B) BARS REPELLED AS CURRENT FLOWS

(C) WHEN CURRENT REVERSES, BARS STILL REPELLED

Now an interesting fact about meters which use the principle of repulsion between two magnetized pieces of iron is that the repulsive force varies as the *square* of the current rather than directly as the current through the coil. Because both bars are magnetized, each contributes to the force which moves the movable bar. If a unit value of current through the magnetizing coil displaces the movable bar .04 inch from the stationary bar, doubling the current in the magnetizing coil will increase the separation between the two bars four-fold, or separate them 0.04 x 4 or 0.16 inch. If the current is tripled the separation increases nine-fold, always changing as the square of the current change.

The above figures of physical displacement are purely illustrative and serve only to correlate the measurement of a varying quantity, such as the sine waveform a-c voltage in Fig. 2-2, with the behavior of the measuring device.

In any application of alternating currents it is useful to express both currents and voltages in terms of their root-mean-square (rms) valuels. Rms values are important because they are the only ones connected with a-c quantities which will give correct answers when used in the power formula $P = I^2R$. Thus, unless there is a special reason for doing otherwise, alternating currents and voltages are always expressed in rms values. The rms value is derived by taking the square root of the average of the squares of all the instantaneous values included in a half-cycle, or any number of complete half-cycles. For the commonly encountered sine wave, the rms value is simply 0.707 times the peak value (see Fig. 2-2). For waves of other forms the rms value has a different relation to peak or average value.

Referring again to the sine waveform curve shown in Fig. 2-2, the fixed relationship between the peak and the rms values make it possible to convert any rms value to a corresponding peak value, and since the positive and negative peaks of a sine waveform are alike in amplitude, differing only in polarity, the maximum amplitude limits of an alternating current (or voltage) are expressed as the peak-to-peak value. Hence any a-c meter which is calibrated in rms values can state the corresponding peak-to-peak values on the same scale. Such multiple calibration is found on some meters, although not too frequently.

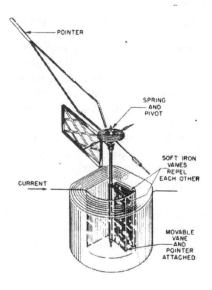

Fig. 2-4. Details of construction of the radial-vane type of moving iron meter.

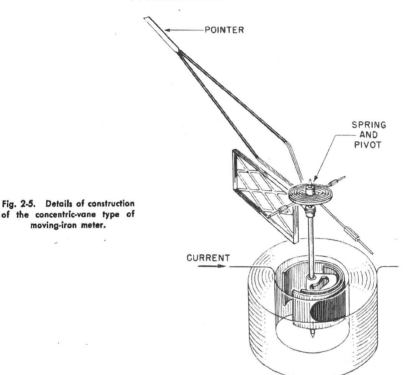

POINTER

SPRING
AND
PIVOT

CURRENT

Fig. 2-5. Details of construction
of the concentric-vane type of
moving-iron meter.

Because of the square relation between the displacement and current, explained above, the displacement of the moving iron in this type of meter is directly proportional to the rms value of the current being measured. Thus this type of meter can read a-c and d-c values on the same scale and is not sensitive to waveform.

The basic principle described in Fig. 2-3 is applied in two groups of a-c meters, the radial vane type and the concentric vane type.

2-3. The Radial and the Concentric Vane Meters

In the radial type, two rectangular vanes are placed inside the coil as illustrated in Fig. 2-4. One vane is fixed and the other is free to rotate on pivots. The movable vane has a pointer attached. When current goes through the coil, the iron vanes become magnetized and repel each other. The movable vane swings away, carrying the pointer across the scale, thus indicating the amount of current flowing through the coil.

In the concentric vane type shown in Fig. 2-5, the soft iron vanes are semicircular, with one inside the other. When the current flows through the coil, lines of force pass through both vanes. The movable vane swings away from the stationary one and moves the pointer across the scale. The amount of swing depends upon the strength of the current in the coil and the control action of the springs.

2-4. Scale Calibration in Moving-Vane Meters

Chapter 1 explained how the moving-coil permanent magnet d-c meter employs a *linear* scale, in which a given difference in current is represented by a fixed distance along any part of the scale (Sec. 1-5). Moving-vane meters, however, have *non-linear* scales. The numbers in the lower portion of the scale are crowded together, while those at the high end are spread further apart. This type of scale is illustrated in Fig. 2-6. It is known as the *square-law* scale and is the type which results

Fig. 2-6. Non-linear, "square-law" meter scale.

whenever the measured current operates through two agencies at the same time. Each magnetized vane is an agency. If a given measured current through the meter is doubled, each vane is magnetized twice as strongly. Since each vane then repels the other twice as much, the combined repulsion between the two vanes is four times as great. In other words, the force of repulsion and the pointer swing do not vary *directly* with the current but vary as the *square* of the current.

2-5. The Hot-Wire Current Meter

This meter uses the physical effects of heat generated by current flow in a wire. The physical change (thermal expansion) in the wire moves a pointer which registers the current in the circuit. Figure 2-7 shows the arrangement. The meter is connected in series with the load by means of terminals A and B. Inside the meter, wire 1 carries the current. Wire 2 is attached to the current carrying wire near the center and pulls on it with a tension controlled by a spring. Wire 2 passes along the round base of a pivoted indicating needle and is used as a drive wire to move the needle. When current passes through wire 1, the wire heats up proportionally to the square of the strength of the current. The higher the current the greater the amount of heat generated, and the more wire 1 expands. As it expands, the drive wire 2 is pulled down by the tension of the spring. Wire 2, riding along the base of the pivoted needle, drives the pointer across the scale. The pointer is adjusted to read zero with zero current by means of a screw regulating the normal sag in wire 1.

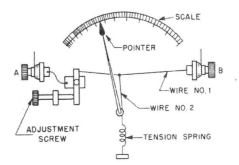

Fig. 2-7 Organization of component parts in "hot wire"
ammeter.

The hot-wire meter is suitable for use not only at power-frequencies but is also useful at radio frequencies. While the hot-wire type is still in use to some extent, the development of thermocouple devices has greatly reduced its application. The relationship between the current and the force developed in the meter to move the pointer is in accordance with the square law, hence the scale used resembles that shown in Fig. 2-6. The weaknesses of this type of instrument are numerous and have been responsible for its replacement by other types of equipment. The zero setting is indefinite and requires frequent readjustment, it is slow acting and it requires the consumption of an appreciable amount of energy in order to cause the generation of sufficient heat, hence it is not as sensitive as other types.

2-6. The Electrodynamometer

The action of a movable coil carrying current in a fixed magnetic field supplied by a permanent magnet was discussed in Chapter 1. (Sec. 1-2, Fig. 1-1.) The action is similar if the permanent magnet is replaced with another coil carrying current. The electromagnet then supplies the fixed field against which the magnetic field of the moving coil acts while rotating to indicate current intensity.

The principle of the electrodynamometer (sometimes shortened to "dynamometer") is illustrated in Fig. 2-8. The schematic symbols of the coils are shown in approximately their operating positions. The turns of the fixed coils are distributed equally on both sides of the movable coil. When current flows through the coils, the polarity of the magnetic fields is such that the movable coil rotates to align itself with the fixed coils. If the current through the coils is reversed, the magnetic polarity of both the movable and the fixed coils is reversed. The *relative* polarities are still the same, so the motion is still the same as before. For this reason, the electrodynamometer is suitable for the measurement of both direct and alternating currents. It is not useful

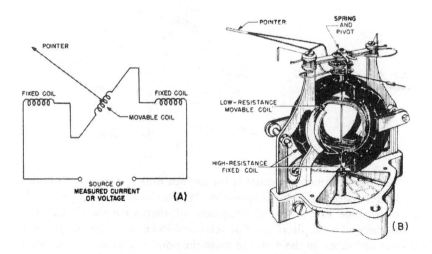

Fig. 2-8. The electrodynamometer: (A) schematic showing how the magnetic fields of the fixed and movable coils interact to cause the latter to rotate; (B) constructional details.

for alternating current of higher than power frequencies because of the mass of the moving parts.

The deflecting force in this type of meter is derived from the current in two ways at the same time; the current acts through the field it creates around the movable coil and again through the field it creates around the fixed coil. For this reason, the scale of this meter follows the *square law* and is like that shown in Fig. 2-6. The electronic technician and experimenter is most interested in the dynamometer because of its use as a *wattmeter*, which is explained in Chapter 4.

2-7. Thermocouple Meters

These meters function by virtue of a difference of potential generated at a junction of two dissimilar metals when current flows through the "heater" element. The voltage so generated is applied across the coil terminals of a moving-coil permanent magnet meter. It is described in detail in Chapter 6.

2-8. Suppressed-Zero Type of Scale

It is often desirable to have a meter read only between particular values of interest rather than have it cover all the values from zero up to the full-scale value; as, for example in the case of reading the a-c line voltage, where we are interested primarily in values between 105 and 125 v. These instruments are known either as *suppressed-zero* types, or sometimes as *expanded-scale* types. One such scale, designed for spreading the full scale to indicate only values between 105 and 125 v ac is shown in Fig. 2-9. This method, while producing maximum readability between the a-c values of interest, requires more complex construction

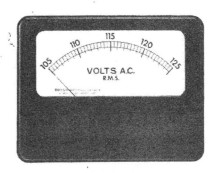

Fig. 2-9. Expanded-scale (suppressed-zero) type of a-c meter. *Beckman/Helipot*

and greater care in calibration, resulting in correspondingly greater over-all cost than the conventional a-c instrument. Details regarding this type of meter are discussed in Chap. 14.

2-9. Classification of Commonly Used A-C Meters

In classifying a-c meters by types, the element of suitability for measuring a-c voltage waveforms at various frequencies is an additional factor that must be taken into account, besides the factor of low or high impedance that was important with d-c meters. The range of frequency desired might be very wide and extends from 60 cycles to well beyond hundreds of megacycles. For convenience, it is well to divide this broad range into three slightly overlapping ranges[1] as follows:

a. Power Frequencies: centering around 60 or 4000 cycles

b. Audio Frequencies: covering a measurement range between 30 cycles up to the limit of audibility (about 20 kc) and, in some cases, extending as high as the supersonic frequencies (about 100 kc)

c. Radio Frequencies: covering a measuring range above 100 kc and extending into the VHF range of hundreds of megacycles, and beyond.

It is necessary to have a wide variety of specific types of a-c meters to provide accurate operation within the frequency ranges needed for a given application. Some of these general-purpose types are discussed in individual chapters later and others are covered in Chap. 14. Table 2-1 provides an initial classification of a-c meter types and their applica-

[1] Video frequencies (from dc up to 4 or 5 mc) straddle all three ranges.

tions; only the most commonly used types are listed to show how they fit into the over-all picture as to their impedance and frequency characteristics.

TABLE 2-1.　CLASSIFICATION OF A-C METERS BY IMPEDANCE AND FREQUENCY RANGE

A-C Meter Type	Impedance	Frequency Range
Moving-iron	Low	Power-frequency range, 50–60 or 380–420 cycles
Electro-dynamometer	Low	Power-frequency range, 50–60 or 380–240 cycles
Rectifier (non-amplifying)	Medium	Audio frequencies
Electronic (VTVM, etc.)	High	Up to high a-f range, without probe; extends into high r-f range with probe
Heating-effect (thermocouple and hot-wire)	Low	Complete range from d-c into the high radio frequencies.

3

ADAPTING METER MOVEMENTS FOR CURRENT AND VOLTAGE MEASUREMENTS

By itself, a meter movement has limited applications. Its resistance and current range may be appropriate for some special uses, more often for current measurement than for voltage measurement. Thus it becomes necessary to examine the basic capabilities of these meter movements as measuring devices, and to see how they may be adapted for a wide variety of practical applications.

3-1. Meters Are Basically Current Operated Devices

All meter movements which involve coils such as have been described are basically current operated devices. The meters function by virtue of the current which flows through the coil portion of the complete mechanism. Thus it would appear from both the d-c and a-c meters we have examined that the meter should be usable as a current measuring device by itself. However, since there is a maximum limitation in the *amount* of current which can be passed through the coil, there is a maximum current indicating range of the instrument when using the meter movement as is.

Based on the physical construction of the moving coil (d-c) and the moving vane (a-c) meter movements, greater limitations exist in the wire size used for the moving coil meter than for the moving vane variety, since the coil rotates in the former and heavy wire would unduly increase weight and inertia. Therefore, it is understandable that, when the coil alone is used without any auxiliary devices, the current measuring range of the d-c moving coil meter is very low. Yet we know that d-c current measurements embrace a great range, from a microampere up to thousands of amperes. How this is done will presently be explained.

The same is true of the a-c moving vane meter. While this type of mechanism permits heavier wire for the current carrying coil, since the coil is fixed in position, it too has limitations. However low and high values of alternating current *are* measured. So it stands to reason that some means is used in order to adapt the meter movements to the wide range of current measurements which must be made. The nature of this means is explained in this chapter.

3-2. Meter Coils Have Resistance

While it is true that the moving-coil and the moving-iron types of meters operate differently, both have one thing in common: both types make use of wire wound coils. They may not be the same in all meters but, whether the coils have few turns or many turns, each pre-sents a certain amount of d-c resistance. For that matter the a-c meter coil also presents inductance to the a-c current, but this is generally so slight at the power frequencies to which these meters are applied that we can neglect the reactance of the coil.

At the moment we are not concerned with how much d-c resistance is present in the meter movement; all that matters is to bear in mind that it exists. The reasons are explained in subsequent paragraphs.

3-3. Meter Coils Are Subject to Voltage Drop

A meter movement presents resistance of some value to the flow of the current, therefore it stands to reason that a voltage drop will develop across the coil — a drop equal to the current times the resistance. This voltage drop is very slight, but this is not the important point. What is significant is that while we speak of the meter as being basically a current operated device, a current meter is also a voltmeter. *It is a voltmeter of such range as corresponds to the maximum IR drop across the coil while the meter is indicating maximum current.*

For example, suppose that we imagine a moving coil meter rated at 1 ma d-c maximum current and a coil (meter) resistance of 100 ohms dc. When the meter is indicating maximum current flow, a voltage drop equal to .001 x 100 or .1 volt (100 millivolts) appears across the meter coil. Hence it is a voltmeter with a maximum range of 100 millivolts, for it takes that much voltage to cause 1 milliampere of current to flow through the meter movement and so indicate full scale.

This situation is not limited to d-c meters only; it is just as true with the moving vane a-c meter. Every a-c current meter is therefore a voltmeter too, yet not in the practical sense. The reasons for this are the conditions which surround voltage measurement. Voltage meas-urements are made *across* the limits of a circuit, and the inherent resistance of the meter movement (dc or ac) is generally so low as to defeat every possibility of using the device as is, as a voltmeter. It must be "adapted" to measure voltage. As far as the meter is concerned, conditions in the instrument are exactly the same whether a voltage is applied or whether a voltage results from the flow of current. However, because the voltage present in a circuit where the full scale meter current flows far exceeds the permissible voltage drop across the meter, auxiliary means must be used to measure the circuit voltage, and yet keep the voltage drop across the meter at the proper value.

Thus while a basic meter movement, either ac or dc, appears in both current and voltage measuring devices, and may be used for the measurement of low values of current without any additions, the use of supplementary devices are required for voltage measurement.

3-4. Adapting Meter Movements for D-C Measurements

Measurable d-c quantities are current, voltage, and power. In this chapter we deal only with current and voltage; power is treated separately in Chapter 5. As to the measuring means, it is the permanent magnet moving-coil type of meter movement. The moving vane a-c meter also is usable for d-c measurements, but it is so seldom employed for this purpose that we ignore it here.

The moving-coil meter movement can be adapted to provide measurement of currents and voltages in excess of those measurable by the movement alone.

3-5. The Basic Principle of the Shunt

The current range of a basic meter movement can be increased by connecting the meter into a circuit in such manner that only a *part* of the total circuit current flows through the meter movement. If the current through the meter bears some known proportionality to the total circuit current, the meter scale can be calibrated to show

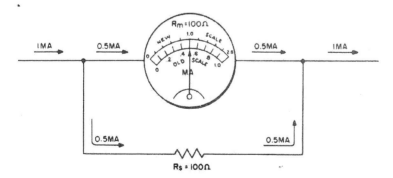

Fig. 3-1. How the current range of a meter is increased by addition of a shunt resistor.

the *total* circuit current corresponding to any value of current flowing through the meter.

In the case of moving-coil d-c meters, this is done by connecting an auxiliary resistor *across* the meter coil to act as a shunt path for some of the current. As in any parallel circuit, the current in the branches is inversely proportional to the resistance of the branches. By suitable design of the *shunt* around the meter, the current through the meter can be made any desired *fraction of the total current* flowing in the circuit where the meter is connected. This is illustrated for a single shunt in Fig. 3-1.

Assume a d-c meter rated at 1 ma maximum and a coil resistance of 100 ohms. If this meter movement is shunted by a 100-ohm resistor, half of the current will flow through the meter and half will flow through the shunt element. Immediately then the meter as a current indicator is capable of showing twice as much current flow as it can without the shunt, since 1 ma current flow in the circuit will result in only 0.5 ma through the meter, hence half-scale deflection. With a 1:1 ratio between the shunt resistance and the meter resistance, full scale indication on this meter would mean twice as much circuit current as is shown on the instrument.

If the shunt resistance is equal to one-half of the meter resistance, twice as much current will flow through the shunt as through the meter, hence the meter will carry only one third of the total current, and its maximum range of current indication has been increased threefold. By making the shunt equal to one-ninth of the meter resistance, nine times as much current will flow through the shunt as through the meter, so that the current indicating ability of the meter has been increased ten-fold.

By arranging a series of shunts which are selectable, the current indicating ability of a meter can be increased a thousand or more times its basic current carrying capacity, all the while keeping the full scale current through the meter within the original design figures. This is commonplace in many commercial meters. By suitable selection of the shunts controlled by a switch or a number of terminals, the current ranges can be multiplied in any desired steps. By proper selection of the multiplying factors, a single current scale serves several ranges, or more than one scale can be used. An example of these is given in Fig. 3-2. The switch must not be operated while current is flowing as this may burn out the meter. This is discussed further in Chapter 8.

Referring again to Fig. 3-1, the illustration of the meter face shows the addition of a new scale with 2-ma full scale indication in place of the original 1 ma. Such modification on a scale is not easy; a

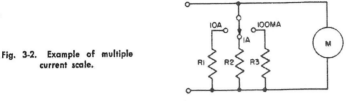

Fig. 3-2. Example of multiple current scale.

much more convenient arrangement is to select such shunts as enable the use of the original meter scale, multiplying the pointer indication by specific amounts as set by the shunt in use.

The practice of fabricating shunts for current meters is not recommended, except perhaps when the amount of resistance required is substantial, as for example an ohm or more. Because of the relatively low resistance of current meters, high multiplying ranges require very low resistance shunts. These necessitate accurate manufacturing processes.

Universal shunts are available as current range multipliers. These are usable with current meters of any coil resistance rating. They consist of a potentiometer arrangement as shown in Fig. 3-3, wherein the current range multiplying factor is a function of the ratio R/R1 and is independent of the meter resistance. If R/R1 is made equal to 1, 30, 100, 300, and 1,000 at the various switch positions, the current scale is increased by exactly the same amounts when the switch is in the respective positions. The initial requirement is that the total resistance R be such that when the switch is in position A (1), the meter read full scale. This can be accomplished by means of a supplementary variable resistance which shunts the whole divider. It is adjusted for full scale meter indication. It will not disturb the relative ratios as the switch position is advanced. The universal shunt is explained more fully in Sec. 8-3 in Chapter 8.

It is interesting to note that most current meters of the moving coil variety contain internal shunts across the moving coil. In other words, the current carrying capacity of the basic meter movement itself is very low, and is increased to the required amount by means of a shunt, different shunts being used for meters of different current rat-

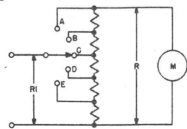

Fig. 3-3. Arrangement of Universal shunt.

$$R_s = \frac{R_m}{N-1}$$

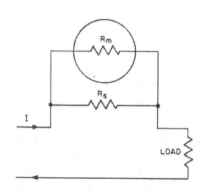

R_m = METER COIL RESISTANCE

N = FACTOR BY WHICH METER RANGE IS TO BE INCREASED

R_s = SHUNT RESISTANCE

Fig. 3-4. Formula for calculating shunt resistors to increase the current range of a basic meter movement.

ings. Generally moving coil meters of a certain general class are constructed alike, and are increased in current range as described above. Moving vane a-c current meters are treated differently, as explained in Chapter 6.

3-6. Calculating Shunts

Although the construction of low resistance shunts for d-c meters is not recommended except when adequate facilities exist, Fig. 3-4 shows the equation used when a shunt is to be calculated. Assume a 50 microampere d-c meter with a coil resistance of 1,000 ohms. What should be the shunt connected across it to make it read 1 ampere full scale? Calculation shows that the current range must be increased 20,000 times, hence the shunt approximates 1/20,000 of 1,000 ohms. Substituting the accurate figures in the equation, it reads:

$$R_m = 1,000 \qquad N = \frac{1}{0.00005} = 20,000$$

$$R_s = \frac{1,000}{20,000 - 1} = 0.05 \text{ ohm}$$

As a convenience, if N is greater than 100 the numeral 1 in the denominator can be omitted, the error thus introduced is negligible for all cases other than when extreme accuracy is required.

3-7. The Basic Principle of the Voltage Multiplier

The voltage drop across a meter movement is fixed by its internal design. To make the device a voltmeter with a range higher than its own internal design establishes, it is necessary to make the instrument a part of a voltage divider system. A portion of the total applied voltage is dropped across the *multiplier* resistance and the remainder of the voltage is dropped across the meter movement. The latter voltage drop is of course never higher than planned for in the design of the meter.

The multiplier circuit is simple. It consists of an external resistor in series with the meter as shown in Fig. 3-5. Assume a meter rated at 1 ma and a coil resistance of 100 ohms. The voltage drop across the movement is 0.001 x 100 or 0.1 volt (100 millivolts). To make this device capable of indicating 1,000 volts full scale requires that 1 milliampere of current flow through the meter when 1,000 volts are applied. Regardless of how much voltage is applied to the system, the voltage across the meter must never exceed the product of the full scale current and the coil resistance, which in this instance is 0.1 volt.

To keep the current at 1 ma with 1,000 volts applied requires that the *multiplier* resistance must be E/I which is 1,000/0.001 or 1,000,000 ohms. However the meter has an internal resistance of 100 ohms, hence the multiplier resistance is 1,000,000 ohms less the 100 ohms meter resistance. Obviously the meter resistance in this case is so slight, relative to the required multiplier resistance, as to be negligible; thus we

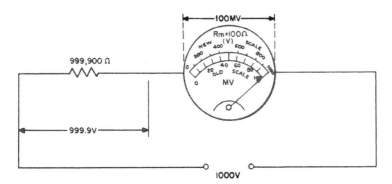

Fig. 3-5. How the voltage range of a meter movement may be increased by the addition of multiplier resistors.

need not concern ourselves with it. However, for the sake of accuracy, let us assume that the multiplier is 999,900 ohms, in which case 999.9 volts will be dropped across the multiplier and 0.1 volt will be dropped across the meter, the sum of these two drops being 1,000 volts. The meter current will be 1 ma, and full scale indication then will correspond to 1,000 volts applied. The meter can then be so calibrated.

With a fixed ratio between the multiplier resistance and the meter resistance, any indication on the meter bears a fixed relation to the applied voltage. With 1 ma indication being the equivalent of 1,000 volts applied, a current indication of .2 ma then would correspond to:

$$\frac{.2}{1} \times 1,000 = 200 \text{ volts}$$

With a direct proportionality existing between the current and the voltage for a certain multiplier resistance, the meter can be calibrated in terms of voltage rather than current, although it must be understood it is always the current through the coil which is causing the movement of the meter pointer. Thus, in Fig. 3-5 the old current scale of 0 to 1 ma is converted to a voltage scale from 0 to 1,000 volts; the .1 ma point on the old current scale corresponds to the 100 volt point on the new scale; the .2 ma point to 200 volts; the .3 ma point to 300 volts, etc., until the full scale position is reached.

In similar manner the use of a series of multipliers can change the 1 ma current meter into a multi-range voltmeter. For example, if the multiplier resistance in Fig. 3-5 is made 4,900 ohms then the total resistance of the system, which is meter resistance plus multiplier resistance, is 5,000 ohms. Since the current is limited to a maximum of 1 ma, the maximum voltage that can be applied is 5,000 x 0.001 or 5 volts, so the meter can be calibrated in voltage from 0 to 5 volts. Using this scale as the basis, the introduction of a different multiplier, say of 499,900 ohms (500,000 ohms in practice) instead of 5,000 ohms, or the addition of 494,900 ohms to the 5,000 ohms, makes the device a voltmeter with a maximum range of 500 volts, since 500,000 x 0.001 $=$ 500 volts. Using the precise value of 499,900 ohms for the multiplier and the 100 ohms in the meter, the total resistance is 500,000 ohms. There is then a 499.9-volt drop across the multiplier and a .1-volt drop across the meter movement.

Modern moving coil d-c voltmeters are developed this way; a current meter is operated in conjunction with multipliers. Three of the most common current ratings for meters used in this manner are: 1 ma, 50 microamperes, and 10 microamperes. Ohm's law shows that the lower the current rating of the meter for full-scale deflection, the higher must be the multiplier resistance for any given applied voltage; first, because the voltage drop is lower across the meter movement, and second, because the current permitted through the meter is less for full scale deflection. In comparison with the 1-ma meter for 1,000 volts applied, which requires a series multiplier of 1 megohm, the 50 microampere meter requires twenty times as much resistance or 20 megohms because the permissible maximum current is one-twentieth.

The use of multipliers in this manner is not limited to d-c moving coil meters only; it can be applied to moving vane meters, although the high current requirements of these meters sets a limit on the suitability of this means of increasing voltage scales. The potential transformer described in Chapter 5 is preferred. Where a-c voltmeters with minimum loading effect are needed, the moving coil meter operated in conjunction with a rectifier is used. This is discussed in Chapter 5.

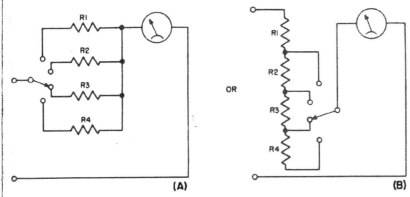

Fig. 3-6. Switch control of multi-range voltmeter amplifier (A) in individual multipliers, (B) series connected multipliers.

Multi-range voltmeters using multipliers are generally arranged as in Fig. 3-6. Each switch position provides a new voltage range and, by suitable selection of the values for *R1*, *R2*, *R3*, and *R4*, one scale may be used for all ranges. For each range, the scale reading is simply multiplied by a factor appropriate to the multiplier for that range.

3-8. Total Voltmeter Resistance

For any one given meter with a fixed current rating, the higher the voltage range arranged for by the multiplier, the higher is the so-called *total meter resistance*. The resistance of the meter movement does not change, but the total resistance presented by the voltmeter and its multipliers increases as the voltage range increases. Thus, for the 1-ma meter previously mentioned, the 5-volt scale means a total resistance of 5,000 ohms; for the 100-volt scale it is 100,000 ohms; for the 500-volt scale it is 500,000 ohms; and for the 1,000-volt scale it is 1,000,000 ohms. For the 50-microampere meter these total resistance values are 100,000 ohms, 2 megohms, 10 megohms, and 20 megohms respectively. For the 10-microampere meter they are 500,000 ohms, 10 megohms, 50 megohms, and 100 megohms respectively.

As in the case of commercial direct current meters which are equipped with internal shunts, the voltmeters available on the commercial market already are equipped with internal multipliers. When they are not switch controlled, they are connected internally to the different voltage rated terminals of the meter. This does not prevent the addition of external multipliers to further increase the range of the instrument. The calculation of the multiplier then is based on the *additional* voltage range required at the rated full-scale meter current. If the maximum range of a voltmeter is 1,000 volts and 2,000 volts is *desired,* the additional voltage drop required across a multiplier is

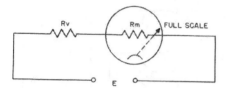

$$R_V = \frac{E}{I_m} - R_m$$

R_V = MULTIPLIER RESISTOR

E = FULL-SCALE VOLTAGE RANGE DESIRED

I_m = FULL-SCALE CURRENT FOR METER ALONE

R_m = METER RESISTANCE

Fig. 3-7. Formula for calculating the proper multiplier resistance for increasing the range of a voltmeter. As explained in the text, in many cases R_m can be neglected in the formula.

1,000. If the meter movement is rated at 1 ma, the external multiplier resistance required in series with the 1,000 volt terminal to raise the voltmeter range to 2,000 volts, is 1,000/0.001 or 1 megohm. For a 50 microampere meter it would be 20 megohms.

3.9. Calculating Multiplier Resistance Values

The equation in Fig. 3-7 enables calculations of the multiplier resistance required to adapt any current meter as a voltmeter. For example, if the meter is a 50-μa instrument rated at a coil resistance of 1,000 ohms, and it is desired to convert it to a voltmeter to indicate 5,000 volts full-scale, the following is a sample calculation:

$$E = 5,000 \text{ v}$$
$$R_m = 1,000 \text{ ohms}$$
$$I_m = 0.00005 \text{ amp}$$

then
$$R_v = \frac{5,000}{0.00005} = 100,000,000 \text{ ohms}$$
$$= 100 \text{ meg}$$

Obviously the meter movement resistance of 1,000 ohms is negligible when the multiplier resistance is so very much greater. If the multiplier resistance is at least 100 times the meter resistance, the latter can be neglected entirely.

The required voltage of 5,000 was deliberately chosen to illustrate that high voltage measurement requires the exercise of great caution if shock is to be avoided. Specially constructed multiplier units are required and must be well insulated and protected against moisture.

3-10. Accuracy and Ratings of Shunts and Multipliers

If the accuracy of a meter movement itself is from 2 to 5 percent, not much is gained by having the shunt or multiplier resistor accuracy better than 1 or 2 percent, since the maximum possible accuracy is that of the meter movement alone. However, if the meter is one of the especially accurate laboratory or comparison standard types, the auxiliary resistors must be precise in their values.

A variety of precision resistors are available for use as multipliers. Some examples are shown in Fig. 3-8. They are available in accuracy ratings of ± 1, $\pm\frac{1}{2}$, $\pm\frac{1}{4}$ and $\pm 1/10$ percent. Various power ratings of from $\frac{1}{2}$ watt to 5 watts are also available.

Shunt and multiplier resistors must not only be accurate in their resistance values, but also must *hold* this accuracy for the life of the instrument with which they are used, in other words, indefinitely. For this reason they are usually ruggedly built and protected against the effects of moisture. If the elements are exposed to humidity the resistance value may be changed slightly. Since some resistors change value with different values of applied voltage, resistors chosen for multiplier use must have very low voltage coefficients so that they will change but negligible amounts during operation.

Fig. 3-8. Examples of precision resistors suitable for use as shunts or multipliers.

 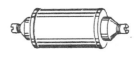

For most voltage measurements in electronic circuits, an accuracy of ± 10 percent is adequate for servicing and general testing. For this reason, when special accuracy is not required, the owner of a meter may use less costly 5 or even 10 percent resistors for the multipliers.

3-11. Power Ratings of Shunts and Multipliers

Power ratings of shunts and multipliers must also be kept in mind. For multipliers, the full-scale current must be within the dissipation rating of any multiplier used. Since power is I^2R, the amount of power dissipated for any voltage range is proportional to the resistance of the multiplier resistor. For example, in the case of the 1-ma meter movement the multiplier resistance should be 100,000 ohms for a 0–100 v scale. This resistor would have to carry 1 ma maximum. The power dissipated by this resistor then would be:

$$P = I^2R = (.001)^2 \times 100,000 = 0.1 \text{ watt}$$

Thus, although the largest voltage scales may require 2-watt resistors, 1 watt sizes usually are sufficient.

For most meter movements the power ratings of shunts are very low when ranges below 10 or more amperes are encountered. For example, it was shown that the required shunt resistance to make a 50-μa, 1,000-ohm meter read 1 amp full scale was 0.05 ohms. The current in the meter is negligible compared to the total current. The power dissipated in the shunt is thus:

$$P = I^2R = (1)^2 \times 0.05 = 0.05 \text{ watt, or 50 milliwatts}$$

3-12. Multipliers and Shunts for A-C Meters

It must be remembered that, if a meter multiplier is carrying alternating current, the *inductance* of the shunt as well as its resistance can be important. Hence it is advisable to use non-inductive elements such as resistors, although random wound, wire wound resistance coils are suitable if the resistance is relatively low, say less than 20,000 ohms. In view of the frequency limits (50—125 cps) imposed on moving-vane types of meters, the ordinary calculations as given for the d-c moving coil meters will be found suitable for these instruments, provided that the meter constants are those for the a-c meter on hand. As to the dynamometer type of instruments, shunts and multipliers should be secured from the original manufacturer of the meter. Further information for increasing the current and voltage ranges of a-c meters, including current and potential transformers, is given in Chapter 6.

It should be noted that the resistances for the lowest voltage ranges (up to 10 v) must be corrected by subtracting the resistance of the meter movement itself, which in these cases is an appreciable percentage of the total resistance.

3-13. Determining Unknown Scale Ranges

Occasionally a useful meter is obtainable, but without any accompanying information concerning it other than that it is dc or ac. To adapt such a meter for use, certain facts must be determined, such as the internal resistance and maximum current range. Both of these factors can quite easily be determinable with a small amount of readily-available equipment. First, let us consider the maximum current range.

To determine the current range of a meter movement, sample currents are passed through it and the deflection noted. Full scale deflection is the desired condition. However, in obtaining full scale deflection, care must be exercised not to start with too much current and so overload, and possibly damage, the meter. Since the full scale current rating may be only a few microamperes, start with zero current and have some way of increasing the current uniformly so that it is always under control. Another current meter which is known to be accurate is required for comparison purposes. This second meter is connected in series with the meter whose scale range is to be determined.

Figure 3-9 shows the test circuit. Connect a 2-watt potentiometer (volume control) across a 1.5-v flashlight cell. A total resistance of anywhere from 50,000 to 100,000 ohms is suitable for R. The arm of the potentiometer (its movable tap) is then connected to the meters in series, then back to the low end of the potentiometer. If desired, a fuse may be added as shown, rated at about 50% over the comparison meter

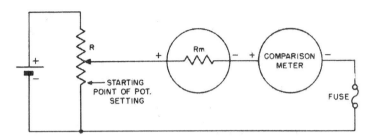

Fig. 3-9. Method of finding meter range by comparison with another meter known to be accurate. Potentiometer R is used to limit the current to within the range of the meters.

full scale reading. *Before the meters are connected be sure the pot is adjusted all the way to zero voltage output.* Then, make sure the polarity of the battery and the meters is consistent throughout, as shown in the figure. The pot arm is slowly advanced· until the meters indicate, then continued until the unknown meter indicates full scale. The full-scale current rating of the unknown meter then equals the current indicated on the comparison meter. The comparison meter can be a current meter from a volt-ohm-milliammeter, and adjusted to different scales as the need arises.

3-14. Determining Meter Resistance

If a meter is of a standard make and model number, information as to current (or voltage) range and internal resistance is easily obtained from the manufacturer. If, however, this information is not available, as in the case of surplus meters, it is possible to determine it. We have already explained how to determine the full scale current range of a meter for which the internal resistance is unknown. Now let us determine the internal resistance of the meter.

A method for measuring the internal resistance is shown in Fig. 3-10. A variable resistor *R1* of value in the order of twice the resistance

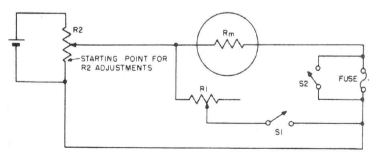

Fig. 3-10. Method for determining meter resistance.

of the meter (this is of course a guess as to the meter resistance) is connected across the meter. A starting value can be 500 ohms. A switch *S1* is connected in series with it to allow it to be removed from the circuit, although this can of course be done by simply disconnecting it. A fuse is added for protection, and must be shorted out for final readings. The meter-*R1* parallel circuit is connected to a potentiometer *R2* (50,000 ohms) which is in turn connected across a flashlight cell (1.5 v) as shown. *R2* must be turned down to give zero voltage (bottom of resistor on this diagram) before the meter is connected.

First, *R1* is removed from the circuit and *R2* is adjusted for full-scale deflection. Then *R2* is left set while *R1* is connected into the circuit and adjusted for half-scale deflection. When this has been done, *R1* has the value equal to the meter resistance, which can be determined by measuring *R1* or reading its calibration if it is a calibrated resistor. If only a small part of *R1* is used, measurement may be critical. This means that *R1* is too high in ohmic value, and a lower value is needed.

The theory behind this method is that when the meter deflection is reduced to half scale, the current must be dividing equally between the meter and *R1*, so *R1* must be equal to meter resistance. R_m and *R1* are so small compared to *R2* that the assumption that the same total current flows with or without *R1* is valid.

3-15. Ohms-Per-Volt Sensitivity

In Chapter 1 it was pointed out how some basic meter movements have different full-scale current ratings than others. Meter movements with the lowest full-scale current ratings are referred to as the most "sensitive," because they respond most to the smallest current values.

When a meter is used as a voltmeter another method is used to designate sensitivity, although it actually indicates the same kind of thing. In this method, a meter is said to have so many "ohms-per-volt." This applies both to the meter movement as well as to its combination with any multiplier resistor.

For example, consider the meter of Fig. 3-5. As shown there, the meter movement alone reads 100 mv full scale, and has a resistance of 100 ohms. The ohms-per-volt is found by dividing the resistance in ohms by the full scale internal voltage drop in volts. Thus:

$$\text{Ohms-per-volt} = \frac{100}{0.1} = 1,000 \text{ ohms per volt}$$

Actually the ohms-per-volt sensitivity depends only upon the full-scale current reading of the instrument alone, and not upon its resistance, as can be seen by trying the above calculation for a 1-ma movement of various other resistances, remembering that the full-scale voltage rating also changes with resistance in each case.

Ohms-per-volt sensitivity also is the reciprocal of the full-scale sensitivity in terms of current. In other words, the full-scale current rating in amperes divided into 1 gives the ohms-per-volt for the meter movement. In the case of the 1-ma meter above, the ohms-per-volt sensitivity is 1 divided by one-thousandth of an ampere. Thus:

$$\text{Ohms-per-volt} = \text{Full scale current in amperes} = \frac{1}{0.001} = 1{,}000$$

·Let's return to Fig. 3-5 and check the other ranges to see that the ohms-per-volt rating of a meter remains the same no matter what multipliers are used. If the full-scale voltage is 1,000 volts and the total resistance is 1,000,000 ohms, the ohms-per-volt rating is 1,000,000/1,000 = 1,000. If the full-scale voltage is 10 volts and the total resistance is 10,000 ohms, the ohms-per-volt rating is 10,000/10 = 1,000.

It becomes clear, then, that no matter what multiplier resistance is used, the same meter movement still has the same ohms-per-volt sensitivity rating. The lower the full-scale current rating of a meter the higher its ohms-per-volt rating.

Why are we so much interested in ohms-per-volt sensitivity? Because if a voltmeter resistance is too low, it may load a high impedance circuit and cause error in the voltage reading. By means of the ohms-per-volt sensitivity it is possible to calculate the total resistance of a voltmeter when set to a given scale, and thus know how likely it is to load the circuit in which the measurements are being made.

3-16. Voltage Measurement

Voltmeters whether d-c or a-c always are connected *across* the circuit, between the two points where the voltage is to be measured. This is shown in Fig. 3-11, and it is to be noted that the kind of voltage applied to the circuit is omitted. The reason for this is that the fundamental requirement stated above applies to all sorts of voltage, and is independent of the nature of the circuit.

Concerning Fig. 3-11, meter 1 measures the total voltage being applied to the circuit; meter 2 measures the voltage drop across R2;

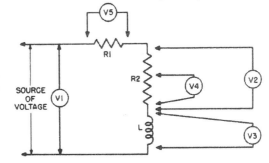

Fig. 3-11. Application of voltmeters for voltage measurement.

meter 3 measures the voltage drop across *L;* meter 4 measures the voltage drop across a portion of *R2* and meter 5 measures the voltage drop
across *R1.* The voltage measurement.across *L* warrants several remarks.
If the applied voltage is a-c, then the voltage measured across *L* is the
IZ drop at the frequency of the applied a-c voltage, which of course
assumes that the voltmeter is capable of response at that frequency. On
the other hand, if the applied voltage is dc the voltage drop across *L* is
the IR drop, where *R* is the d-c resistance of the winding and I is the
current through the winding.

Whether or not a voltmeter is capable of measuring the IZ drop
across a circuit element subjected to an a-c voltage depends entirely on
the design of the measuring device. Suitable instruments are discussed
in this book, but in the meantime a very pertinent detail concerning
voltage measurement warrants comment. This is the ohms-per-volt rating of the voltmeter relative to its loading effect on the circuit under test.

3-17. Loading Effect of the Voltmeter Resistance

The effect of voltmeter resistance on a high-resistance circuit may
be appreciated from a study of Fig. 3-12A, 3-12B and 3-12C. In Fig.
3-12A, the resistors *R1* and *R2* are shown shunted across a 100-volt battery. Since *R1* and *R2* are equal in resistance, the voltage divides equally
across them; and since the total voltage is 100, the voltage across each
resistor is 50 volts.

When a voltmeter is connected across *R2,* as shown in Fig. 3-12B,
the circuit then becomes effectively that shown in Fig. 3-12C, in
which the voltmeter is represented as the resistance R_m. Assume that
we are using a d-c voltmeter with a rating of 1,000 ohms-per-volt on the
100-volt scale, its resistance is 100,000 ohms. Since R_m is in parallel with
R2, and $R_m = R2,$ the effective resistance when the meter is across
R2 then becomes 50,000 ohms, just one-half that without the meter.

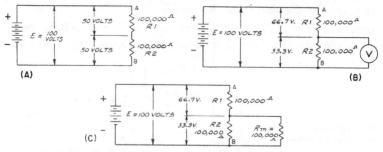

Fig. 3-12. Effect of voltmeter resistance on circuit resistance: (A) with voltmeter not
present; (B) with voltmeter connected across R2; (C) equivalent circuit of B with voltage
connected across R2 shows how the voltage drops across R1 and R2 are changed from
those shown without the voltmeter.

The total circuit resistance between A and B is now no longer 200,000 ohms, but 150,000 ohms; and the voltage across $R2$ and the meter will be equal to the ratio of 50,000 ohms (representing the resistance of $R2$ and the meter in parallel) to the total circuit resistance of 150,000 ohms. Therefore, only one-third the total voltage E across the circuit will be indicated by the meter. So, instead of being 50 volts, the *loading* effect of the meter has reduced the voltage to one-third of 100 volts or $33\frac{1}{3}$ volts. This means that the error in reading amounts to 16-2/3 volts, which corresponds to $33\frac{1}{3}$ percent less than the true voltage when the meter is disconnected from the circuit.

It is easily seen from this illustration that the error in reading due to loading effect in high-resistance circuits can be far greater than meter inaccuracies due to manufacturing causes. For this reason it is imperative to use a voltmeter of high resistance rating in high-resistance circuits.

Suppose for the moment that the meter has an ohms-per-volt rating of 20,000, in which case the 100-volt scale presents a total meter resistance of 2,000,000 ohms. Placing this resistance across $R2$ in Fig. 3-12A, results in a final value of

$$\frac{100,000 \ + \ 2,000,000}{100,000 \ \times \ 2,000,000} = 95,200 \text{ ohms}$$

The total circuit resistance then becomes $R1 \ + \ (R2$ in parallel with $R_m)$ or 195,200 ohms. The voltage across $R2$ then is

$$\frac{95,200}{195,200} \times 100 = 48.7 \text{ volts}$$

or within 3 percent of the correct value. The advantage of the higher ohms-per-volt rating is evident. This is even more important as the circuit resistance becomes higher and the circuit voltage becomes lower.

Although the examples cited in Fig. 3-12 are based on d-c voltages, the same condition applies to a-c circuits; the voltmeter resistance must be high in comparison with the resistance of the circuit across which the voltage measurement is made. As to what is meant by a high ratio between voltmeter and circuit resistance, a suitable minimum is about 20:1, preferably higher. As much as it may be desired this cannot always be realized, in which event the error may be more than a few percent and will have to be tolerated. The important thing is to realize the possibility of circuit loading when the voltmeter is applied. The subject also receives attention elsewhere in this book.

It is in the area of sensitivity that the permanent-magnet moving-coil meter movement shows its great superiority over a-c/d-c movements, such as moving-iron types and dynamometers, for measurements in electronic circuits. The sensitivity of these latter movements is relatively very low, ranging from 1 to 300 ohms per volt, depending upon the range and type. The sensitivity is lowest in these instruments for the lowest voltage ranges, a 2-v meter sometimes drawing as much as 0.5 amp for full-scale deflection. Obviously, the usefulness of these meters in electronic circuits is limited to measurement of power line voltage and current, filament (heater) voltage and current, and the like. Of course the dynamometer is very useful in the form of a watt-meter for servicing radio and tv receivers, communication receivers and other devices, as explained in Chapter 4.

3-18. Current Measurements

Whether ac or dc, when current measurements are made, one significant condition must be borne in mind. Irrespective of the type of current meter used, the instrument always must be connected in series with the circuit where the current is to be measured. In this way the current flowing in the circuit will flow through the meter. This is shown in Fig. 3-13, where M is the meter and R is the load. The fact

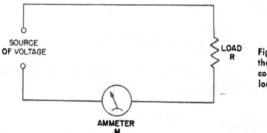

Fig. 3-13. For current readings, the meter should always be connected in series with the load whose current is to be indicated, as shown here.

that the load is shown as a resistance does not limit the circuit arrangement to resistances. The load may take any form: it may be a lamp or a vacuum-tube filament; it may be a d-c or a-c motor; it may be the plate, screen, or cathode circuit of a vacuum tube; it may be a transformer winding; in fact it may be anything through which current flows. The type of meter used is, of course, determined by the character of the current, its magnitude, its waveform and the accuracy of measurement required. Very low-current meters with high movement resistance are used in a similar manner. The high resistance of the meter seldom disturbs the circuit to which it is added because low-current circuits usually are of very high resistance.

In the event that the circuit has a number of branches, measurement of the current in each of the branches calls for the use of a meter *in each branch;* and measurement of the total current requires the use of a meter in that conductor of the complete circuit which is common to all the individual branches. A simple example of this is shown in

Fig. 3-14 wherein meter *M1* measures the total current, meter *M2* measures the current flow in the *R1* branch and meter *M3* indicates the current in the branch containing the series combination of *L* and *R2*.

One point, or one lead, in the circuit in which measurements are to be made is nearly always grounded, although this need not be so. Various potentials may then exist at other points in the circuit with respect to ground. If a current is to be measured, the *ammeter should be inserted at the lowest potential point* with respect to ground.

For example, in Fig. 3-14, if the lower side of the source of voltage is grounded as shown, the total current should be read with the meter in the position *M1* rather than in the lead from the top of the voltage source, (shown by dotted line meter *M1*), where the potential is high. Similarly, in the branch circuits, *M2* and *M3* are properly connected as shown by the solid line drawings instead of above *R1* and *R2* respectively, as shown by the dotted line drawings.

3-19. Effect of Adding Current Meter to Circuit

Adding a current meter in series with a circuit is the equivalent of adding resistance and inductance in series with the circuit. In a d-c circuit the inductance of the meter coil is unimportant. In the a-c circuit-operated power frequencies, the same can be said to be true. However, in both cases, the resistance of the meter coil can be important because it tends to reduce the current in the circuit being measured. Whether the amount of reduction is significant depends entirely upon the original circuit resistance without the meter in place, and the circuit resistance with the meter connected into the circuit.

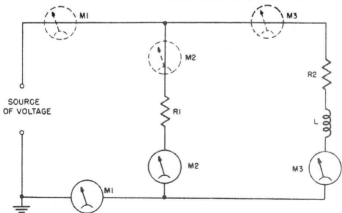

Fig. 3-14. This circuit shows how an ammeter can be connected in series with *any* branch of a circuit to indicate the current in a *particular* branch, or in the supply lead for the *total* current in the branches. Meter M1 measures the total current; M2 measures the current through R1, and M3 measures the current through R2 and L.

Fig. 3-15. This formula is used
to find how much the ammeter
will read low due to the resistance of the meter in the
circuit.

$$\text{PERCENT DECREASE IN CURRENT} = \frac{R_m}{R_c + R_m} \times 100$$

$R_m =$ METER RESISTANCE

$R_c =$ CIRCUIT RESISTANCE WITHOUT METER

Moving-coil d-c meters present low coil resistance whereas the moving-vane a-c meters present higher values of resistance. However, if the original circuit resistance, without the meter in place to measure the current, is at least 100 times the meter movement resistance, this effect of the meter is negligible. Reducing the ratio to 50 makes the measured current about 2 percent lower than the true current without the meter in the circuit. This much difference is not too high, but it is possible that the difference may become appreciable depending on the circuit resistance and the meter resistance. The equation shown in Fig. 3-15 enables determination of the percentage decrease in current indication due to the presence of the meter. For example, if the circuit resistance is 2 ohms and the meter resistance is 0.05 ohm, the indicated current will be low by 2.4 percent:

$$\frac{0.5}{2+0.05} \times 100 = 2.4\%$$

4

ADAPTING METER MOVEMENT FOR RESISTANCE MEASUREMENT

4-1. Resistance Measurement; Voltmeter-Ammeter Method

One of the most direct approaches to resistance measurement, and perhaps the simplest in principle, is simultaneous measurement of both the voltage across and the current through a circuit or component. We know from Ohm's law that resistance equals voltage divided by current, and thus can calculate resistance from the voltage and current readings taken.

The circuit for the voltmeter-ammeter resistance measurement method is illustrated in Fig. 4-1. It is seldom used except by students studying basic electronics, because the ohmmeter is abundantly available as a resistance measuring device.

For accurate results in measuring resistance by the voltmeter-ammeter method, the relative resistances of the voltmeter and ammeter themselves must be kept in mind. For low values of R (comparable with the internal resistance of the ammeter A) the resistance R_m must be deducted from the value of the circuit resistance, determined by dividing the voltage V by the current indicated on A. For higher values of R, the circuit shown is satisfactory without worrying about R_m.

4-2. The Ohmmeter

The ohmmeter is an instrument that measures the d-c resistance of a circuit element or network, indicating the value of resistance on a meter scale calibrated in ohms. It is also used to locate open circuits and to check circuit continuity, and can be used to provide a rough check on capacitors. In application, the ohmmeter is one of the fundamental test instruments, along with the voltmeter and ammeter.

Basically, the ohmmeter consists of a sensitive current meter, a source of low d-c voltage, and some form of current-limiting resistor.

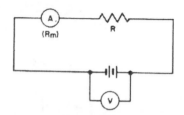

**Fig. 4-1. Circuit for the volt-
meter-ammeter method of measuring resistance.**

The meter itself is usually of the conventional moving-coil permanent-magnet type, and a battery supplies the d-c voltage.

The values of d-c resistance encountered in electronic equipment vary from fractions of an ohm to many millions of ohms (megohms). Consequently, most ohmmeters are constructed to cover a number of ranges from very low to very high resistance. This is accomplished by connecting various values of current-limiting resistors into the ohmmeter circuit. The individual ranges can be selected as required by means of individual input terminals or a selector switch. Multi-range ohmmeters are discussed further in Chapter 8.

4-3. Series-Type Ohmmeter

Figure 4-2 illustrates the circuit of the basic series-type ohmmeter. A 4.5-v battery and a variable resistor, R_A, are connected in series with a milliammeter. The two leads labeled *P1* and *P2* represent test prods which are connected across the resistance to be measured, R_x. The meter is 0—1 ma movement, requiring 1 ma of current for full-scale deflection. The internal resistance of the meter, R_M, is 50 ohms. The variable resistor R_A adjusts the series resistance in the circuit so that 1 ma full-scale current flows when the test prods are shorted together. This occurs when the *total* series resistance is 4,500 ohms (4.5 v/4,500 ohms = 1 ma). In the circuit shown, the meter resistance (50 ohms) and R_A (4,450 ohms) make up the required series resistance of 4,500 ohms.

When the test prods are shorted together, current flows through the circuit and the meter pointer is deflected across the scale, usually to the right. The resistor R_A is then adjusted to provide full-scale deflection of the pointer. This position of the pointer corresponds to zero external resistance, since the two connected prods represent a short circuit across the ohmmeter terminals. In the series-type ohmmeter full-scale deflection of the meter pointer indicates the lowest resistance; the opposite end of the scale represents the highest resistance. A knob on the front of the ohmmeter marked *zero-ohms-adjust* controls R_A.

In most cases R_A consists of two resistors, one fixed and one variable, whose total value is equal to, or slightly more than, the value required for R_A. As the battery ages its terminal voltage decreases. Consequently, the variable zero-ohms resistor portion of R_A compensates for this decrease. The current can be increased to the required value by an adjustment of this variable resistor which decreases the total series resistance. The zero-adjusting resistor R_A can be connected either in series or shunt with the meter to provide the same function. When it is connected in shunt, a fixed resistor is connected in series with R_A, and both resistors are connected across the meter. Another fixed resistor is then required in the series-ohmmeter circuit to limit the current to the approximate required value. The final current is adjusted by R_A. For accuracy in meter readings, a zero-ohms adjustment must be made each time the ohmmeter is used.

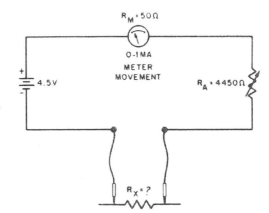

Fig. 4-2.   Circuit of simple series
type ohmmeter.

After the ohmmeter has been adjusted for full-scale deflection, the test prods are separated and the meter-pointer returns to the open-circuit position on the left of the scale. Placing the prods across the unknown resistor connects the latter in series with the ohmmeter circuit. As a result the current is reduced proportionately, and the meter no longer deflects full-scale. If the value of R_x is equal to the combined resistance of the current-limiting resistor R_A and the internal resistance of the meter, or 4,500 ohms, the total circuit resistance then becomes ,500 + 4,500, or 9,000 ohms. The current in the circuit now becomes

$$I = \frac{E}{R} = \frac{4.5}{9,000} = 0.0005 \text{ amp} = 0.5 \text{ ma}$$

Thus the pointer is now deflected only to the half-scale point, which for this ohmmeter is calibrated to indicate 4,500 ohms. If the value of another unknown resistor should be twice as great as the circuit resistance of the ohmmeter, then the total resistance is tripled and the current is reduced to one-third the full-scale deflection value. Accordingly, the meter pointer deflects to one-third full-scale. This point represents a measured resistance of 9,000 ohms.

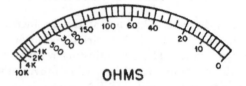

Fig. 4-3. Typical meter scale
for series type of ohmmeter.

OHMS

In this way the remainder of the meter scale is calibrated for ohmmeter use. A typical series-type ohmmeter scale is shown in Fig. 4-3. Notice how crowded the divisions become at the high-resistance (low current) end of the scale. For this reason, readings should always be taken in the higher-current (lower resistance) three-quarters of the scale, and the scale (or range) used should be chosen to make this possible.

4-4. Shunt-Type Ohmmeter

For reading very low values of resistance, the *shunt-type* ohmmeter circuit is better than the series-type just described. The shunt-type ohmmeter circuit is shown in Fig. 4-4. The unknown resistance R_x is now shunted across the meter, instead of in series with it. With the unknown resistance connected in this manner, some of the current in the ohmmeter circuit takes the path through R_x. The current through the meter is reduced accordingly, and the amount of deflection drops proportionately. The current passing through the meter depends upon the ratio of the shunt resistance of R_x to the internal resistance of the meter.

In operation, the current in the shunt-ohmmeter circuit is first adjusted by the zero-ohms adjustment R_4 to provide full-scale deflection on the meter. In making this adjustment, the test prods $P1$ and $P2$ are *not* shorted to each other, as in the case of the series-type ohmmeter, but are kept apart. The total resistance in the ohmmeter circuit is the sum of R_4 (4,450 ohms) and the meter resistance (50 ohms), or 4,500 ohms. A current of 1 ma flows through the ohmmeter circuit, giving full-scale deflection on the meter.

After the zero-ohms adjustment, the test prods are connected across the unknown resistor R_x, placing it in shunt with the meter. The 1-ma current in the ohmmeter circuit divides at the junction of the meter and R_x, part of it passing through R_x and the rest flowing through the meter. The reduced current through the meter, which is proportional to the resistance of R_x, causes a reduced deflection of the meter pointer.

Note that for the shunt-type ohmmeter the greater the unknown resistance being measured the greater is the current through the meter. For this reason the scale is such that the indicated resistance increases progressively from *left to right*, right being maximum for full-scale

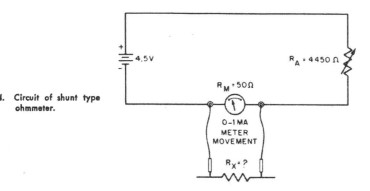

g. 4-4. Circuit of shunt type ohmmeter.

eflection of the meter. A typical shunt-type ohmmeter scale is shown n Fig. 4-5. Notice that the direction of increase of resistance is just he opposite of that for the series-type ohmmeter scale shown in Fig. 4-4.

4-5. Voltage-Divider or Potentiometer-Type Ohmmeter

With the series-type ohmmeter the scale tends to be overcrowded near the high resistance end. To reduce this crowding, and distribute the scale divisions a little more evenly, another type of circuit which uses the voltage-divider principle is used. The circuit is shown in Fig. 4-6. The meter is a high-resistance voltmeter and is connected across R_s in series with R_A, so it reads the voltage across R_s although it is calibrated in ohms. R_s is connected with the unknown resistance R_x so that the two form a voltage-divider across the battery. The voltage across R_s, indicated by the meter, depends upon how much resistance appears across the test prods. When the prods are shorted together the voltage across R_s is maximum, being equal to the full voltage of the' battery, and R_A is adjusted for full-scale meter deflection. The higher the unknown resistance R_x, the lower the voltage across R_s and the lower the meter reading in current units. Thus, as is the case in the series-type ohmmeter, the highest resistance indications are at the left end of the scale. However, the scale divisions are somewhat more uniform than those of the series type.

4-6. High-Resistances and the Megger

Some instruments are available for measuring very high resistance values. They operate on the ohmmeter principle; the applied voltage is increased to a relatively high value so measurable currents can be produced through high resistances. These devices are known as "meggers."

The megger is a portable insulation-resistance test set for measuring extremely high values of electrical resistance, up to 100,000

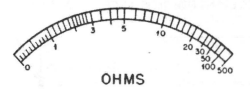

OHMS

Fig. 4-5. Typical meter scale for
shunt type ohmmeter.

megohms or more. Such high resistances are encountered during checking of the insulation resistance of cables, resistance between conductors

of multiple cables, between windings and from winding to ground in
transformers, between windings and from windings to ground in rotating equipment, and paper, mica, and ceramic dielectric capacitor
leakage.

A typical megger and carrying case is shown in Fig. 4-7, the schematic diagram is Fig. 4-8. When the crank is operated, the d-c generator applies 500 v with the polarity indicated. Electron current flows
from the negative terminal of the generator, through the potential
coil, (which forms part of the ohmmeter movement), through the
large resistor R2, and back to the positive terminal of the generator.

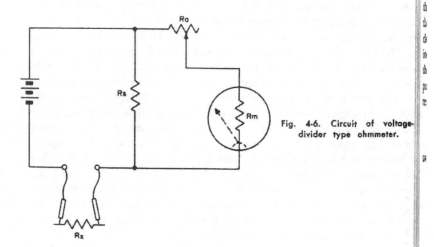

Fig. 4-6. Circuit of voltage-
divider type ohmmeter.

The magnetic field, set up by the potential coil, reacts with the field
set up by the permanent magnet in such a way that the coil and
pointer assembly move in a counterclockwise direction. With nothing
connected between the terminals LINE and GROUND, the only current
flowing is the current just described. As a result, the pointer indicates
infinite resistance on the scale.

Fig. 4-7. A typical megger.

When an unknown resistance is connected between the LINE and GROUND terminals, the electron path is from the negative terminal of the generator, through the current coil (part of the ohmmeter movement), through resistor *R1*, out of the LINE terminal, through the unknown resistance, into the GROUND terminal, then to the positive side of the generator. If the unknown resistance is extremely low, most of the current flows through the current coil and the pointer moves close to zero on the scale. If the value of the unknown resistance is increased, less current flows through the current coil with respect to the fixed amount of current in the potential coil. Consequently, the pointer moves from zero toward infinity on the scale and indicates the resistance of the element or system connected between line and ground.

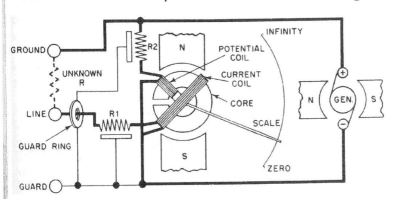

Fig. 4-8. Circuit of a typical megger. The metal guard ring and associated components prevent leakage currents along the surface of the instrument case when readings over 100 megohms are being made.

POWER MEASUREMENT AND WATTMETERS

5-1. Measuring Power in D-C Circuits

In d-c circuits, and in a-c circuits where the components are pure resistances, power is equal to the current times the voltage and thus may be determined by measurements of voltage and current in the circuit under test. Since voltage and current are related by Ohm's law to resistance, power may also be determined by measurements of resistance and either current or voltage. The various power relationships which exist in d-c circuits are expressed in Fig. 5-1.

Thus, in Fig. 5-2, if the voltmeter reads 12 volts and the ammeter indicates a current of 1 ampere, the power in the load circuit, R_L, is 12 x 1 or 12 watts. If we know that the resistance of the load is 12 ohms and the current is 1 ampere, then the power is 1^2 x 12 or 12 watts. If the voltage is 12 volts and the resistance is 12 ohms, the power is $12^2/12 = 144/12 = 12$ watts.

It seems quite simple to connect a voltmeter across a circuit and an ammeter in series with a circuit to determine both the current and voltage in the circuit. However, there are certain considerations which must be observed, in making these connections, otherwise the measurements may be considerably in error.

Figure 5-2 illustrates the above statement. If the resistance of the voltmeter is high its manner of connection is not of much importance; but since it is possible that the voltmeter resistance might be low and its current consumption appreciable, mention must be made of the fact that if the voltmeter is connected into the circuit as shown in Fig. 5-2 that the load current as indicated by the ammeter is the *sum* of the current drawn by the voltmeter and the load resistance.

Fig. 5-1. The power in a circuit can be calculated by knowing the resistance and
either the current or voltage, in accordance with the formula on the right.

$$WATTS = E \times I = \frac{E^2}{R} = I^2 \times R$$

E = VOLTAGE IN VOLTS
I = CURRENT IN AMPERES
R = RESISTANCE IN OHMS

This load current would be greater than that found in the circuit of Fig. 5-2 for identical meters and load resistance, and would represent an error in power consumption indication.

5-2. Measuring Power in A-C Circuits

In a-c circuits, power is not always equal to "voltage times current." It is equal to voltage times current only when the circuit is purely *resistive* and there is no reactance present. When reactance is present (inductance or capacitance) it influences the current or voltage but *absorbs no power*. A pure reactance alternately stores energy and returns it to the circuit; the amount absorbed equals the amount given up so that the consumed power is zero. In this case (pure reactance) the voltage and current are 90° out of phase with each other. If (as is more common) there is some reactance and some resistance in the circuit, the power consumed is less than voltage times current (*EI*) but is more than zero. If the *power factor* of the circuit is known, the power can be determined from the voltage and current readings by the following formula:

$$P = EI \ (pf)$$

where P = actual power consumed
  E = a-c voltage across the load
  I = alternating current through the load
  pf = power factor

The power factor expresses the relative amount of resistance in the circuit compared to total impedance. Thus

$$pf = \frac{R}{Z} = \cos \theta$$

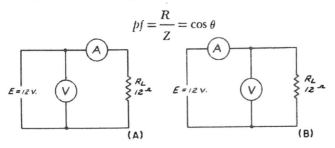

(A)   (B)

Fig. 5-2 If the ammeter A is connected as shown in A at the left, only the current drawn by the load R_L will be indicated on it; if it is connected as in B, right, then the ammeter reading will be the current drawn by both the load and the voltmeter, V.

where $R =$ circuit resistance

$Z =$ total circuit impedance

$\theta =$ phase angle between current and voltage

In order to measure accurately the power in *any* a-c circuit, we must have some instrument which takes into account the power factor as well as the current and voltage. Such an instrument is the *wattmeter.*

However, in practice, many power readings are simplified because: (a) the voltage is reasonably constant, and (b) the power factor is reasonably close to 1.0. This is true of most electrical appliances in the home, where the power line voltage is standardized at 117 v and 60 cps. Such appliances as ranges, heating pads, broilers, heaters, coffee-makers and the like are simply pure resistors connected across the line. Motor-driven devices, such as vacuum cleaners, hair-dryers, pumps, sewing machines etc., have a power factor only slightly less than 1. Power transformers used in television and radio receivers have a power factor between .9 and .95.

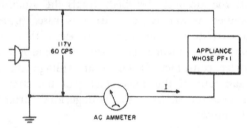

Fig. 5-3. Diagram of arrangement which measures power with an ammeter providing the voltage is constant and the power factor of the load is 1. (1 = 100%).

POWER WATTS·117 x I

Because of constant known voltage and unity power factor, satisfactory approximations of power readings can be made simply by connecting an a-c ammeter in series with the grounded lead, as shown in Fig. 5-3. The power is then equal to the voltage (117 v in this case) times the current (as indicated by the current meter). Sometimes ammeters are calibrated in terms of power, assuming a specified voltage and unity power factor. As long as these conditions exist, and to the degree which they exist, these power meters are quite accurate.

In many locations or equipments, the available voltage may fluctuate enough to make the ammeter method inaccurate. In many homes, especially in rural areas, power line voltage may change as much as 25 percent or more. Obviously, in such cases, the voltage at the moment of measurement must be known.

Fortunately, there is an instrument which takes into account all factors, current, voltage, and power factor at the same time, and indicates on one scale the power being consumed. This is the electrodynamometer wattmeter, which will now be described.

The theory of the dynamometer-type of wattmeter has been explained on the basis that current flows through a moving coil that is located in a magnetic field supplied by a fixed (voltage) coil. This basis was used in order to highlight the fact that this method for
· producing the deflection torque is similar to the method used in the D'Arsonval moving-coil meter, where again the torque is produced by current flowing through a moving coil located in a magnetic field produced in the latter case by a field magnet. However, where relatively large currents are involved in the measurement of power, the construction of commercial dynamometers is more likely to choose the current coil as the fixed one and to make the voltage coil movable, in order to avoid the difficulty in conducting large currents into and out of the spring suspension of the moving coil. In any case, no matter which coil is the moving one, the terminals to the current and voltage coils are always clearly indicated, and the connections to these terminals are made in the same manner as those shown in Fig. 5-4.

It should be emphasized at this point that measurement of power may be made without necessarily using a wattmeter. The ammeter method explained above is perfectly satisfactory in the frequently encountered cases where the voltage supply is constant and the power factor is unity or close to it. Also, particularly in the measurement of

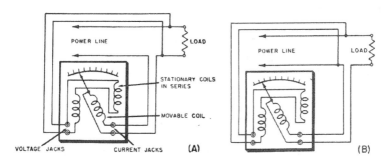

Fig. 5-4. Organization of dynamometer type wattmeter.

audio-frequency output power, the method commonly employed is that of obtaining a voltage reading across a known resistance. For example, if a 4-ohm resistor is used in place of the voice coil of the loud speaker, and an a-c voltage reading of 6 v (at say, 400 cycles) is obtained across this resistor, then the output power (P_0) is E^2/R or $36/4 = 9$ watts for the a-f power output in this case. This method of measuring power output is covered in later sections on meter applications.

In short, where only the product $E \times I$ is desired (unity power factor, or close to it), the wattage can be satisfactorily obtained quite easily by determining either E or I alone, providing the other factor is reasonably constant.

5-3. The Electrodynamometer Wattmeter

In Chapter 2 the electrodynamometer ammeter and voltmeter were discussed, and their basic principle of operation explained. This same meter becomes a wattmeter if the coils are connected a little differently.

The dynamometer wattmeter consists of two coils, one stationary and one moving, just as in the dynamometer ammeter. The stationary coil has many turns of small wire with high resistance. The movable coil has a few turns of large wire with low resistance. It is pivoted and has jeweled bearings, springs for control, and air damping such as the dynamometer ammeter and voltmeter have.

The high-resistance coils are connected across the voltage source, or *across* the device whose power consumption is to be measured. The movable coil is connected in *series* with the device. These connections are shown in Fig. 5-4 (A). Current through the load also passes through the movable coil. Current through the voltage coil is proportional to the voltage across the load. The interaction of the magnetic fields from the fixed and movable coils causes the movable coil to rotate. The torque on the movable coil is proportional to the current and also to the voltage, and is therefore proportional to their *product*. The meter pointer thus registers according to the power consumption, or $E \times I$.

The torque is greatest when the voltage and current peaks occur at the same time (in phase) and becomes relatively less as the current and voltage peaks are gradually separated, in time up to 90 degrees. Thus the instrument also takes into account the phase and power factor.

The movable current coil has low resistance and inductance so as not to disturb the circuit into which it is connected. The inductance of the fixed voltage coil also is low relative to its resistance so the current through the coil is in phase with the voltage. Sometimes a non-inductive resistor is connected in series with the voltage coil to reduce the relative reactance to resistance in the circuit and to keep the current as closely as possible in phase with the voltage.

Since both the current and voltage coils have resistance, the watt-meter itself consumes some power. This loss is usually specified on the instrument, and represents the loss at full scale deflection. A typical wattmeter has a power loss of 2 watts in the voltage coil and 1 watt in the current coil. Since the percentage error of readings due to meter power-consumption tends to be greater for lower readings, the accuracy is best if the upper part of the scale is used.

There are two possible ways of connecting the wattmeter, these are shown at (A) and (B) of Fig. 5-4. At (A), the current through the potential (voltage) coil also goes through the current coil. Therefore the current coil is measuring both the load current and the potential coil current. At (B) the potential coil is measuring the voltage of the load added to the voltage across the current coil. In each case some error is inevitably introduced, because the voltage and current are not exactly those of the load.

Some wattmeters are compensated so that the power loss due to wattmeter connections is corrected for in the instrument. Otherwise, for exact measurements, the power consumed by the wattmeter should be subtracted from the final wattmeter indication. The meter power loss can be easily checked by disconnecting the load. The potential coil is then in series with the current coil. The uncompensated meter reads its own power loss. If the wattmeter is compensated the reading is zero, if there is a numerical reading the number is the correction factor which should be subtracted from the reading when the load is connected.

Both the voltage and current coils have definite ratings and overload of either can damage the meter. For example, a typical wattmeter is designed to measure a maximum of 200 watts; the current coil has a maximum rating of 3 amperes, and the potential coil has a maximum rating of 130 volts. If either of these ratings are exceeded overload may occur.

In dynamometer wattmeters deflection is linear and practically uniform over the full range of the wattmeter; doubling the power produces twice the deflection. Because the voltage and current coils are separately connected, power is doubled when current through *either* the voltage or current coil doubles. When the currents through *both* coils double there is *four* times the power, and thus four times the deflection.

Dynamometer wattmeters automatically take into account the power factor of the circuit. The wattmeter always reads true power.

5-4. Usefulness of the Wattmeter in Servicing Equipment

The wattmeter is useful not only in determining how much power a device is taking under *normal* conditions, but is also useful as a detector of defective conditions. If some component of a unit is causing leakage or undue loading, the unit draws more power from the line than it normally would. This is particularly true of such units as radio and tv receivers, amplifiers, intercommunication systems, radio transmitters and the like, which are a-c or d-c power-line operated. For example, if a device is rated at 50 watts for normal operation and is consuming 100 watts, there is a definite indication of a defect causing overload somewhere in the circuit. The technician would then look for such things as shorted or leaky filter or bypass capacitors, gassy tubes, short-circuit between tube elements, failure of bias on one or more tubes, etc. These would cause excessive currents to flow and thus dissipate extra power, all of which is supplied by the line, and thus shows on the wattmeter.

5-5. Polyphase Circuits

In measuring voltage (and power) in two-phase and three-phase circuits, care must be taken in interpreting the readings obtained, even though the actual procedure for any one voltage or current measurement is no different than it is in a single-phase circuit. The points to be reviewed here will be limited to interpreting voltmeter readings—these being typical of the considerations that apply in polyphase circuits.

5-6. Two-Phase Circuits

Two-phase circuits, while not as commonplace in power generation as the three-phase systems, are widely used in the fractional horsepower field—particularly in the case of servo motors. The typical servo motor operates by the application of two voltages, 90° out of phase with respect to each other. Generally each phase is electrically independent and has two wires for its connection, resulting in four terminals for the motor, two for the reference phase and two for the control

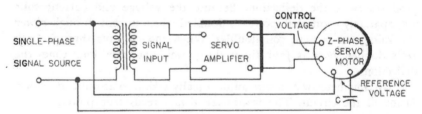

Fig. 5-5. Two-phase arrangement applied to a servo motor; reference voltage is shifted 90° out of phase to the control voltage by capacitor C.

phase, that is, in quadrature with it. In the servo amplifier, these two-phase voltages are usually obtained originally from a single-phase source by using a capacitor in series with one of the voltages to produce the required 90°-phase shift, as illustrated in Fig. 5-5. However, since each of the two voltages is maintained electrically independent of the other, the procedure for interpreting each voltage reading is kept quite straightforward by simply relying on the designations "reference voltage" and "control voltage" to keep each phase distinct from the other.

5-7.   Three-Phase Circuits

The three-phase generator produces three phases, each phase being 120° out-of-phase with reference to the succeeding one. In effect, each phase may be considered as a separate generator, with each of the three phase windings interconnected to produce the over-all generator. There are two methods for interconnecting the individual windings for each

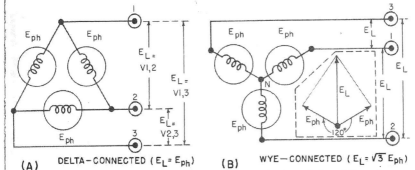

(A)   DELTA–CONNECTED ($E_L = E_{ph}$)   (B)   WYE–CONNECTED ($E_L = \sqrt{3}\,E_{ph}$)

Fig. 5-6.   Three-phase, three-wire generator connections showing relation of line voltage E_L to phase voltage E_{PH}: (A) delta-connected; (B) wye-connected.

phase, resulting in either a delta-connected or wye-connected generator, as illustrated in Fig. 5-6 A and B. In each generator, the line voltage E_L is defined as the voltage existing between any two lines, $V_{1,2}$ and $V_{2,3}$ and $V_{1,3}$. The phase voltage E_{PH} is, by definition, the voltage generated in each phase winding. In the case of the delta-connected generator, seen in Fig. 5-6A, the line voltage E_L is obviously the same as the phase voltage E_{PH}, or

$$E_L = E_{PH} \qquad \text{(delta generator)}$$

However, in the case of the wye-connected generator in Fig. 5-6B, any line voltage E_L is the result of the vector addition of two phase voltages E_{PH}. As a result of the phase conditions prevailing, the vector addition of any two phase voltages results in a line voltage E_L that is $\sqrt{3}$ times as large as either phase voltage E_{PH}, or

$$E_L = \sqrt{3}\,E_{PH} \qquad \text{(wye generator)}$$

By keeping in mind that the voltage obtained between any two terminals of either type of generator is by definition the line voltage, there should be no confusion up to this point, since each generator is a three-phase, three-wire source.

The possibility of misinterpreting voltage readings arises when the wye-connected generator is used with a fourth wire (neutral), connected to point N, as in Fig. 5-7. Here the 208-volt, three-phase, four-wire generator provides a line voltage $E_L = 208$ volts, when measured between any pair of line terminals — 1, 2, or 3. However, when the voltmeter is connected between any of these line terminals and neutral to obtain the line-to-neutral measurements, the voltmeter reads a phase voltage of 120 volts. This relation can be checked from the expression for the wye-connected generator:

$$E_L = \sqrt{3}\, E_{PH}$$

Solving for E_{PH} which is the line-to-neutral measurement,

$$E_{PH} = 208/1.73 = 120 \text{ volts.}$$

It will be noted that the use of this commonly used three-phase, four-wire system makes it possible to obtain 120-volt, single-phase power by simply making a connection between any of the 1, 2, or 3 line terminals and neutral. Thus, the four-wire system can be used either to

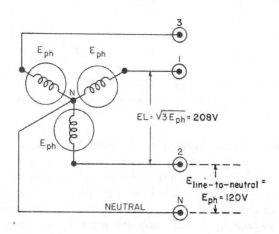

EL = $\sqrt{3}\,E_{ph}$ = 208V

$E_{line-to-neutral}$ = E_{ph} = 120V

Fig. 5-7. Three-phase four-wire generator connection. The line-to-neutral connection is the same as the phase voltage (E_{PH}) which is $E_L/\sqrt{3}$, in this case 120 volts.

supply a three-phase 208-volt load by making a connection to the three line wires, or it can also supply a single-phase 120-volt load by connecting the load between any line and neutral. Although the question of balacing the loads is important, it cannot be considered here. (The reader will find it fully discussed in any text on a-c fundamentals.) The pertinent question of what voltage readings to expect can be completed by noting that the voltage between any two line terminals is 208 volts and that, therefore, this three-phase, four-wire generator can also be used to supply a single-phase 208-volt load, by connecting to any pair of lines. Discussion as to the methods of achieving a balanced system is not necessary.

In summary ,the voltage ratings of three-phase sources are specified in terms of rms line values. Thus, any 208-volt, three-phase supply has 208 volts between lines. Similarly, the voltage rating of three-phase loads (regardless of the type of load connection) is specified in the same effective line values. The power rating of a three-phase load refers to the total power dissipated in the unit.

In measuring three-phase power, individual wattmeters are not generally used because of possible difficulties of connection, especially where the neutral terminal of a wye-connected device might not be available for connection to an individual voltage terminal. Instead, the total power in a three-phase circuit, is usually read from a *polyphase wattmeter*. In such a power meter, the algebraic summation is accomplished internally in the meter and the reading of the deflection obtained on the polyphase wattmeter is therefore a function of the total power. The power rating of a three-phase load is also given in terms of the total power dissipated in the unit.

6

METERS ADAPTED FOR AUDIO AND POWER FREQUENCIES

Some meter movements will read alternating voltage and current as they are and thus do not have to be "adapted." These are the moving-iron or vane and dynamometer types described in Chapter 2. The scales are calibrated in rms values, and sine waveform voltage or current is assumed. However, as has been mentioned, these movements have such low sensitivity that they are of limited use in electronic circuits. Most of the a-c measurements in electronic and electrical circuits in general are now made with a d-c moving-coil permanent-magnet movement, operating in conjunction with some kind of rectifier arrangement which converts ac to dc for application to the meter. It is this adaptation of the d-c permanent-magnet meter with which we are primarily concerned in this chapter. The reason being that rectifier type instruments afford a-c meters of relatively high ohms-per-volt —— from 1,000 to 5,000 ohms-per-volt.

6-1. Adding a Rectifier to a D-C Meter

Modern instrument rectifiers for power frequencies and audio frequencies are mainly of the copper-oxide type. These are constructed of a series of copper discs clamped flat together in a stack. On one side of each disc is a coating of copper oxide forming a layer of copper oxide between each adjacent pair of discs. Washers made of lead are clamped against each oxide surface to improve the contact, and the whole assembly is held tightly together by an insulated bolt running through the center. Typical examples of instrument rectifiers are shown in Fig. 6-1.

The rectifier used with an a-c meter may be a single unit or a group of individual rectifiers connected together. The simplest arrange-

Fig. 6-1. Typical copper oxide instrument rectifiers.

ment is the *single-unit half-wave* rectifier of Fig. 6-2. The rectifier stops currént flow during the negative alternations, leaving the current waveform as shown at the right. One cycle of the meter-current waveform is analyzed in the lower part of the illustration, showing how the average value in the half-wave rectifier system is equivalent to 0.45 times the rms value of the alternating current which would result without the rectifier. This means the rms value is 2.22 times the average value indicated on the d-c scale of the meter. Thus the d-c scale of the meter can be changed to read rms ac by multiplying each value by 2.22.

The copper-oxide rectifier is not a perfect rectifier, but allows a very small back-current tó flow. Thus a slight error is introduced in the arrangement of Fig. 6-2. To minimize back-current, the arrangement of Fig. 6-3 is often employed. Here a second rectifier is connected

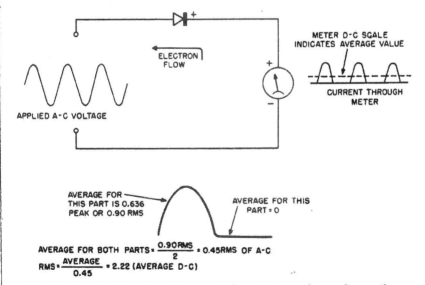

Fig. 6-2. The simplest meter-rectifier circuit. The meter responds according to the average value of the rectified current. In the lower part of the illustration this average is shown to be 0.45 times the rms value of the measured a-c.

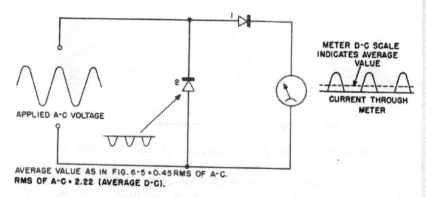

Fig. 6-3. Double-rectifier arrangement, which minimizes "back-current."

across the source of measured voltage so that it draws current when the first rectifier is non-conducting. This shunts the meter and first rectifier during such periods, minimizing the reverse current. The waveform of current through the meter is the same as for Fig. 6-2, and the same relations hold.

The most sensitive arrangement is that using the *bridge* rectifier illustrated in Fig. 6-4. Four rectifier units are connected to each other, to the meter, and to the source in the manner shown. When the source of measured voltage has one polarity one set of rectifiers conducts, and when the source has the opposite polarity the other pair conducts.

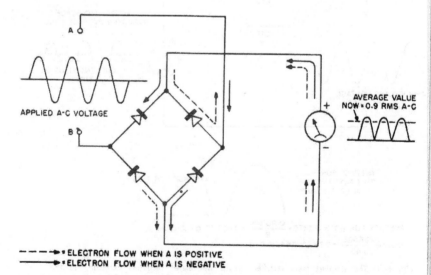

Fig. 6-4. Four section rectifier connected in bridge circuit giving full-wave rectification.

The arrows show the direction of electron flow for each polarity of the source. The result is that current flows in the meter in the same direction for *both half-cycles* of the a-c wave. For this reason, the average value of the pulsating current through the meter is 0.9 of the a-c rms value. This is twice the relative value of the average in the half-wave rectifier circuits of Figs. 6-2 and 6-3. Thus the meter response is twice as great for a given current, and the sensitivity is that much better.

Rectifier units are available in very compact form to take care of all of these types of circuits. The ones with several rectifier units are built with these units stacked together in one physical assembly as shown in Fig. 6-1. Instrument rectifiers have ratings ranging from 1 to 5-v rms and from 1 to 75-ma dc.

Copper-oxide rectifiers are subject to certain limitations which affect their application and the accuracy of the readings. The characteristics of the rectifier change with temperature and condition of overload. The resistance of the rectifier also changes with a change of current through the rectifier. This latter effect is of major importance at low rectifier current densities, (when low values of current are flowing through the meter). In this connection the "forward" resistance increases at low values of current. Therefore the scale of the rectifier-type meter designed for the measurement of low a-c quantities is crowded near the low end. Further, because of this condition, calibration for alternating current does not hold for the measurement of direct current. At high values the rectifier type meter is comparatively linear and provides a substantially uniform scale.

6-2. Problems in Adding Multipliers to the Rectifier Meter

In order to provide practical a-c voltage scales, multiplier resistors must be used in series with the meter movement and the rectifier. As soon as this is done the efficiency of the copper-oxide rectifier is reduced considerably, unless measures are taken to counteract the effect.

This problem is largely overcome by use of the circuit of Fig. 6-3, shown again with a multiplier resistor R in Fig. 6-5. On one alternation current goes through the meter movement and rectifier 1, with the usual process of half-wave rectification taking place; on the other half-cycle current is bypassed around the meter movement and rectifier 1, and is shunted through rectifier 2. The meter and rectifier 1, with 50,050 ohms, are practically an open circuit compared with the 1,000-ohm forward resistance of rectifier 2, and thus draw practically no reverse current.

6-3. Scale-Linearizing Shunt

It has previously been mentioned that the forward resistance of the copper-oxide rectifier increases when relatively low currents are

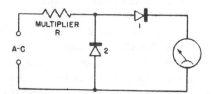

Fig. 6-5. Use of multiplier resistance in rectifier type a-c voltmeter.

being rectified. Forward resistance can vary from 2,000 to 500 ohms as current in the rectifier varies from 0.1 ma to 1 ma. This tends to make the calibration non-linear, with considerable crowding toward the low end of the scale.

This difficulty is partially relieved by shunting the meter, and not the rectifier, with a resistor (as shown in Fig. 6-6). This causes the rectifier 1 to carry more current than the meter; thus, when the meter is reading low values, the rectifier is carrying a high enough current to keep it in the relatively linear portion of its characteristic. Of course the shunt lowers the sensitivity on voltmeter scales, because it takes more current to produce a given deflection. Typically, before the shunt is connected, the sensitivity of the meter in Fig. 6-6 alone is $1/.0002 = 5,000$ ohms per volt; after the shunt is connected it takes 1 ma to deflect full scale, thus the sensitivity drops to $1/.001 = 1,000$ ohms per volt. However, the improved scale calibration is generally considered worth the price of the lost sensitivity, and many commercial instruments use the shunt.

6-4. Sensitivity of Rectifier-Type Instruments

The shunt resistor described previously is placed in the circuit only when a-c quantities are being measured. This is one reason that rectifier-meter sensitivities are ordinarily considerably lower than d-c meter sensitivities in the same combined instrument. Also, the rectifier has a definite amount of forward resistance and impedance which, for simplicity, we have ignored in previous discussions. However these are present, and result in less current through the meter than would be the case in perfectly resistanceless rectification.

Because of these two factors, meter-rectifier sensitivities are ordinarily much lower than d-c meter sensitivities in combination instruments using the same meter movement. For the better instruments commonly available, a d-c sensitivity of 20,00 ohms per volt is common. For a-c voltage measurements, most instruments feature a sensitivity of 1,000 ohms per volt, with a few having as high as 5,000 or 10,000 ohms per volt.

6-5. Frequency Characteristics of Copper-Oxide Meter Rectifiers

Because the copper-oxide rectifier has internal capacitance, bypassing of higher frequencies takes place and the rectified current passing through the meter is accordingly decreased at the higher frequencies. This occurs to a much lesser degree today than in the designs of years ago because meter rectifiers have undergone substantial improvement in design. Presently, the usual frequency error is about 1 percent up to about 20,000 cps, and a copper-oxide rectifier type of a-c voltmeter is down approximately 3 db at around 70,000 cps.

Except for the necessity for correction because of the frequency characteristic of the rectifier, the meters used especially for a-f measurements are the same as those already described. An a-f voltmeter is simply an a-c meter, of the rectifier type, used at higher frequencies. However, the scale of the audio-frequency meter is usually calibrated

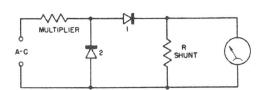

Fig. 6-6. Circuit with scale linearizing shunt resistor.

in different units, either *decibels* (db) or *volume units* (VU). Because of this, such meters are often referred to as "db" or "VU" meters.

6-6. The Decibel and the DB Meter

The decibel or db is a unit of power *ratio*, although it is just as often used in connection with ratios of voltage and current. It was developed to provide a convenient expression for power or voltage ratios of large magnitude, such as are used in audio work and with audio equipment.

The db meter, on the other hand, is a direct reading instrument which is calibrated in db values above and below a reference standard, thereby enabling direct measurement of a power or voltage level relative to the standard.

6-7. The Decibel Reference Level

The decibel is defined by the following equations:

$$\text{Number of db} = 10 \text{ Log } \frac{P2}{P1}$$

$$db = 20 \text{ Log } \frac{E2}{E1}$$

$$db = 20 \text{ Log } \frac{I2}{I1}$$

where *P2* is output power   *E1* is input voltage
    P1 is input power   *I2* is output current
    E2 is output voltage   *I1* is input current

Suppose, in the expression for db in terms of power levels, we substitute a definite standard specified power level for P1, to which we compare any other power level P2. Then, instead of expressing a power level in watts or milliwatts, we could say it is equal to so many ·db, all the while knowing that we mean this many db above (or below) the specified standard reference level.

In order that the standard reference level be expressed in terms of voltage as well as power, the resistance of the standard load must also be specified.

Two different reference levels are commonly used. Unfortunately, neither one has been universally accepted in preference to the other, so the standard used must be specified for each meter. The two standards are shown in Fig. 6-7: the tendency is now toward more general use of the 1-mw standard shown in (B). The units dbm and Volume Units (VU) are the same as db except that the reference level is 1 mw into 600 ohms instead of 6 mw into 500 ohms.

A db, dbm, or VU meter is simply a rectifier type a-c meter with the scale changed from voltage to "level" units. It will be noted that in multimeters in which there are both a-c voltage and db scales, the "0-db" point will correspond to either 1.732v a-c (standard A in Fig. 6-7) or 0.775v a-c (standard B). See Fig. 6-8(A) for the db scale. The figures for db (or dbm or VU) above this zero point are marked with plus signs; the figures below this point are marked with minus signs. Minus quantities on the db scale do not mean any change in polarity of electrical quantities. When the db reading becomes negative, it means that the ratio of the measured voltage to the reference voltage is less than 1. Thus, in a load of 600 ohms, a voltage of 0.775v rms equals O dbm or VU; voltages higher than 0.775v rms have plus values and those lower than 0.775v rms have minus values.

In ordinary multimeters and db meters the rectifier-type a-c meter is used, without any special refinements. For applications in which broadcast program material and other rigidly-controlled audio signals are to be measured, a special meter called the "VU Meter" is used. The reference level, reference deflection on the scale, damping, frequency response, types of scale and other factors are carefully specified. The standard, recommended VU meter scales are shown in Fig. 6-8B. Besides the VU scale, another scale, proportional to (but not equal to) voltage is also included, with an arbitrary value of 100 at the 0 VU point.

VU meters use full-wave rectifiers, usually of the bridge type, because of the following requirement: The indication must be independent of the poling of the indicator when measuring unsymmetrical waves. Such characteristics can be obtained by the use of a direct current instrument in conjunction with a full-wave rectifier.

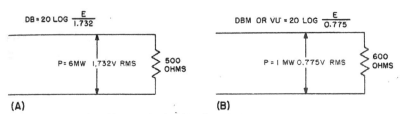

Fig. 6-7. Reference standards for db and VU meters.

In most applications, VU and other level meters are used for *relative* readings, and the zero level may be any voltage desirable or convenient at the moment. If an absolute reading is required, however, it must be taken across an impedance of 500 or 600 ohms, whichever the scale calls for.

6-8. Output Meters

An *output meter* is a db or VU meter equipped with a load resistor across which level readings may be taken. The scale of an output meter is usually calibrated in both db (or VU) and power (watts, milliwatts, microwatts). The load resistor in the meter assembly is used as a test load for the output circuit of a radio receiver or audio amplifier. The power output of the device can then be read directly from the meter scale.

6-9. Effect of Waveform on Rectifier Meter Readings

When a d-c meter is used with a rectifier, it responds to the *average* value of the a-c half-cycles. However, the scale is calibrated in *rms* values. With a sine wave the rms value is 1.11 times the average value (average value times form factor). Therefore the rms calibrations are always 1.11 times the average value which would be indicated at any point.

However, as soon as we try to read a-c voltage of a waveform other than the sine, we run into an error. This is because the form factor

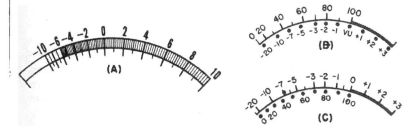

Fig. 6-8. Db scale and standard VU meter scales. The scale at (B) emphasizes arbitrary units proportional to voltage. The scale at (C) emphasizes the VU calibration.

for waveforms other than the sine is not 1.11. For example, Fig. 6-9 shows the familiar "sawtooth" waveform used in the deflection circuits of tv receivers, and its average and rms values. Since the form factor is 1.15 instead of 1.11, the regular rectifier a-c meter reads about 4 percent low. Although the average value operates the meter, its scale is calibrated to indicate 1.11 times the average. Thus the indicated rms value is lower than the actual rms voltage, which is 1.15 times the average value. Of course, if we know what the waveform is, we can correct each reading by multiplying it by 1.15/1.11, or 1.036.

6-10. Current and Potential Transformers

Although we are primarily concerned with problems involving the rectifier type of meter, the basic a-c types, such as the moving-iron and dynamometer, are also encountered. A word about the extension of their scales should be said. Shunts are not as frequently used with these meters, instead current scales are extended by means of *current transformers*.

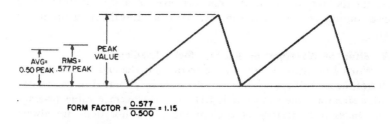

Fig. 6-9. Sawtooth waveform and relative value of the rms and average compared to peak. The form factor here is 1.15, compared with 1.11 for the sine wave.

The secondary winding of the current transformer is connected to the meter, and the primary of the transformer is connected into the circuit where the current is to be measured (as shown in Fig. 6-10). Due to the design of the current transformer, the load current is stepped down by a predetermined factor so that the secondary current is within the operating range of the current meter. The step-down ratio between the primary and secondary windings then becomes a multiplying ratio applied to the current indication of the meter. For example, in Fig. 6-10, the connections to the transformer are such that a step-down ratio of 10 is being used, full-scale deflection of 1 ampere in the secondary circuit is the equivalent of 1 x 10 or 10 amperes flowing in the primary or load circuit.

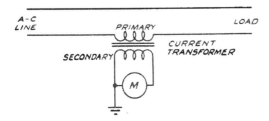

Fig. 6-10. Use of the current transformer. This is the method used to extend the scales of current meters such as the moving-iron and dynamometer types.

Such transformers are designed to cover a wide range of step-down ratios, from as low as 2 times to as high as several thousand times. Usually the transformers are designed to operate with a meter rated at from .5 to 1 ampere full scale, although some small current transformers are designed for operation with 2-ampere meters. These transformers are suitable for use with all types of a-c current meters.

The second method makes use of the potential transformer. The wiring connections to a voltmeter are shown in Fig. 6-11. The transformer is designed to provide a predetermined step-down ratio when used in connection with certain specific meters. These meter specifications state the full-scale range of the meter and its current consumption. In other words, the secondary of the potential transformer must work into a certain definite load. The step-down ratios which cover a wide variety then become the multiplying factors which are applied to the voltage indication upon the voltmeter. Thus, if the step-down ratio is 100 to 1 and the voltmeter indicates 10 volts, the voltage present across the primary of the transformer is 10 x 100 or 1,000 volts. Such trans-formers are usable over a frequency range of from 25 to about 133 cycles.

One precautionary measure which must be remembered when potential transformers are used is that one side of the secondary should always be grounded, as shown in Fig. 6-11.

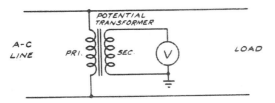

Fig. 6-11. The scales of moving-iron and dynamometer voltmeters can be extended by the use of a potential transformer, connected as shown here. Voltages far in excess of the voltmeter range can be read due to the step-down properties of the transformer.

6-11. Logarithmic-Scale Indication

The decibel scale previously described is a familiar example of a logarithmic scale. The reason for its wide use in sound-level measurements arises out of the fact that the response of the human ear is analogous to a logarithmic variation. Expressed in a different way, the response of the ear to a sound 100 times as great is not linearly proportional to the ratio of 100 : 1 (or 100 times as great), but instead is proportional to the logarithm of that ratio, that is, to the log

Fig. 6-12. Logarithmic scale. Ballantine a-c electronic voltmeter
model 300

100 : 1 (or only twice as great). Since the unit for this logarithm, the bel, is very large, the smaller unit of 0.1 bel, called a *decibel* (abbreviated db) is the unit usually used. By the proper change in constants, it can also be used for voltage or current ratios in addition to the power ratio previously mentioned. And so the logarithmic scale has many important uses outside of the sound-level field.

In general, the logarithmic type of indication is desirable for all cases where the ratio of two measurements is a very large number. Since the over-all gain in amplifiers, for example, often runs up over 1,000 : 1, it is more convenient to express this amount of voltage amplification as + 60 db, as compared to a reference, rather than trying to express the voltage as 1,000 times as great as the input voltage.

The reason the logarithmic scale is able to display much greater ratios on a single range of the meter scale than an absolute voltage scale can is made evident in the illustration of the logarithmic scale in Fig. 6-12. A logarithmic scale covering three decades, is often used in radiation detection meters. By this means it is possible to display tens, hundreds, and thousands of counts per minute (cpm) on a single-scale range.

It will be noted in Fig. 6-12 that the logarithmic type of scale does not have a left-hand zero for zero volts, since, by its logarithmic nature, the db scale readings have meaning only as a measure of a ratio and not as an absolute value. Consequently, when the db scale is combined with a voltage scale, as it is in Fig. 6-12, the initial scale marking for volts is not zero, but some arbitrary value (as 1 v in the photograph). It therefore follows that in its unenergized position, the indicating needle must ordinarily rest to the left of the lowest reading (against its left-hand stop), and no attempt should be made to mechanically move the needle to align it with the first marking on the scale. For any mechanical alignment that may become necessary, the manufacturer's instructions should be carefully followed for this type of meter.

6-12. Clamp-on Type A-C Ammeter

A very convenient method for measuring a-c amperage is available in the form of the *clamp-on* type meter. It makes it possible to obtain a-c current measurements without breaking the circuit. This type of meter is illustrated in Figs. 6-13 and 6-14, which show how the clamp jaws are actuated by pushing the plunger at the left. The a-c ampere

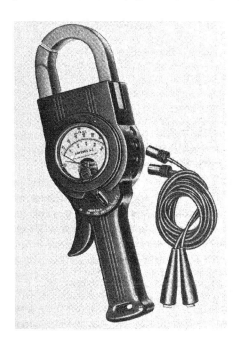
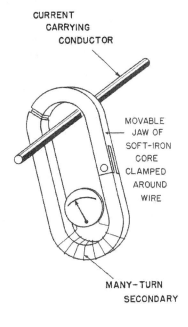

CURRENT
CARRYING
CONDUCTOR

MOVABLE
JAW OF
SOFT-IRON
CORE
CLAMPED
AROUND
WIRE

MANY-TURN
SECONDARY

(B)

Fig. 6-13. "Clamp-on" type ammeter: (A) Multi-scale volt-meter, Weston; (B) schematic representation of the transformer action of clamp-on ammeter with the wire carrying the measured current acting as the primary.

ranges are selected by the switch shown, to enable range selection. Ranges from 6 to 300 amp a-c in six steps are available in the model showed in Fig. 6-14, when the clamp-on adapter is plugged in to the companion volt-ohm milliammeter.

When the clamp-on device is clipped around a single wire carrying alternating current, it operates on the transformer principle, with the split transformer yoke acting as the core for a variable-turn secondary, in which voltage is induced by the varying magnetic field around the single straight wire acting as a primary. Thus the split transformer yoke functions as a magnetic core which completely encircles the straight wire when the pressure on the plunger is released. The voltage induced

Fig. 6-14.   Clamp-on adapter for a-c current measurement with multimeter. *Triplett*

in the secondary then causes a current to flow in the companion meter circuit, whose scale is calibrated to read directly in amperes.

It will be noted that this method, while highly convenient, is limited to the measurement of fairly heavy alternating currents because of the necessity of transferring the effects of the primary current by means of the magnetic core to the secondary. However, it is possible to increase the ammeter sensitivity somewhat by the use of adapters which plug into the line. These are also called *line-separators*, since they allow access to one side of a twin-cord conductor without physically disturbing the cord.

A clamp-on adapter which allows the clamp-on feature to be used with a regular service type of multimeter, is shown in Fig. 6-14. As will be seen ,the lowest full-scale range on the meter selector switch, when used as is, reads 6 amp. By use of the special *line-separator* adapter, the sensitivity of the combination can be extended down to 0.3 amp full-scale.

6-13. Video-Frequency Measurements

The suitability of meters for measuring signals in the video-frequency (v-f) band, including a range of frequencies between the audio (a-f) and radio (r-f) frequencies, is not as sharply defined as it is for the counterpart instruments in either the a-f or r-f band. As stated in Sec. 6-5, the rectifier-type moving-coil meter using a copper-oxide rectifier begins to introduce errors above 1% at around 20 kc, even though deflections up to 70% of normal (3 db down) are still available up to around 70 kc. It is best, however, not to rely upon this type of rectifier meter for measurement of *absolute values* of a-c signals beyond the strictly audible ranges (around 15 to 20 kc). This caution applies equally as well to the volt-ohm milliammeters (see Chap. 8) which employ this type of meter. If only *relative deflections* are desired, this rectifier-type of meter may be used at any one frequency at which a reasonably good deflection is obtained, provided the measurements being compared are all made at this one same frequency. Otherwise, the r-f meters, discussed in Chap. 7, or the electronic type of meter (i.e., the vacuum-tube or transistor voltmeter) discussed in Chap. 9, should be used. This is also true for measurement of pulses and other complex waveforms containing harmonics, as discussed at the end of Chap. 9.

7

METERS ADAPTED FOR R-F MEASUREMENTS

Chapter 6 explained that rectifier type a-c meters are satisfactory for use with frequencies up to about 70,000 cps. Of course, with correction, they can be used much higher. However, they are not suitable for measurement of r-f currents at several megacycles and higher. At these frequencies, special measures are taken to adapt simple meter movements, primarily the permanent-magnet moving-coil type, to current, voltage, and power measurements. There are two ways of doing this: (1) by use of a thermocouple with a meter movement, and (2) by use of a high-frequency tube or crystal-rectifier probe and a meter movement. These methods are considered in this chapter.

7-1. The Thermocouple Ammeter

In radio transmitters, thermocouple ammeters are generally used for measuring r-f current. These meters may be connected in the antenna circuit so as to measure antenna current and power, in a "dummy load" used for testing, or in other parts of the transmitter circuit where r-f current is to be measured. They are also used in induction heating and diathermy units.

A thermocouple consists of two wires, or other pieces, each of a different metal, joined at one end. If the junction of these metals is heated, a d-c voltage appears across the free terminals. This phenomenon is called *thermoelectricity*.

In thermocouple ammeters, use is made of this property of a thermocouple to indicate the temperature of a wire. Since the wire whose temperature is indicated is carrying the current to be measured, the thermocouple voltage is also proportional to the current in the wire. This is because the current in the wire has a heating effect (I^2R) which raises the wire's temperature according to the amount of current

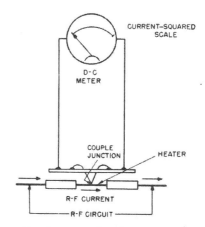

Fig. 7-1. Basic principle of the thermocouple meter. R-f current through the heater generates heat which raises the temperature of the thermocouple junction. The increased temperature of the couple causes it to produce a voltage, deflecting the pointer of the d-c instrument accordingly.

in it. The temperature of this wire applies heat to the thermocouple, which develops a voltage. This voltage is applied to a sensitive meter of the permanent-magnet moving-coil type. Thus the reading of this meter is proportional to the current in the wire. The principle of the thermocouple ammeter is illustrated in Fig. 7-1.

7-2. Thermo-Ammeter Frequency Response and Heater Design

Two factors influence the type of heater used; these factors are frequency and current value. High values of either or both make special physical design of the heater necessary. This is because r-f currents do not distribute themselves uniformly through the cross-section of the conductor. They flow mostly in a thin layer, or "skin" at the outer surface; this effect is known as "skin effect." The higher the frequency, the thinner the skin in which most of the current flows. Since the current flows through less metal at higher frequencies, the effective resistance is higher. The heat generated to operate the thermocouple is determined by I^2R; thus as the effective R increases with frequency, the heat generated for a given current increases. In this manner the thermocouple voltage is increased and the meter reads high at higher frequencies. This effect extends in greater degree to lower frequencies when the current is high, because the heat is proportional to the current squared.

For relatively low frequencies, and currents up to about 2 or 3 amp, a solid wire heater is usually used. Above these values, a *tubular* heater is substituted for the solid one. The thinner the walls of the tube, the less the frequency error. Of course there is a limit to how thin such a tube can be made; this limit is about 1 mil (one thousandth of an inch). The tube is made of a platinum alloy of resistance about 17 times that of copper. With such a heater, frequency error can be kept to about 5 percent up to about 100 mc, and about 10 percent up to 150 mc. 150 mc is about the top limit for which this type of meter should be used for making absolute readings, that is, readings that are to be used in calculating power, etc. However the thermocouple ammeter responds well and can be used as a qualitative indicator, or for comparative readings to see which of two currents is greater, at much higher frequencies.

7-3. Scale Calibration of Thermocouple Ammeters

In these meters, deflection is proportional to the amount of heat generated in the heater wire. This, in turn, is proportional to the square of the current passing through it. Therefore, the thermocouple meter has a square-law scale. Since the calibration divisions in the lower or extreme left, portion of a square-law scale are crowded, readings in that area are less accurate than those taken elsewhere on the scale. Thus thermal meters usually are selected so that the expected value of measured current will place the pointer at least half-way up the full scale range. For example, if the estimated current to be indicated is 7 amperes, the meter range should not exceed 10 amperes.

REGULAR SQUARE-LAW SCALE   (A)

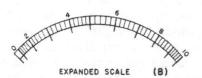

EXPANDED SCALE   (B)

Fig. 7-2. How the scale of a thermocouple meter can be "expanded" to be more uniform over the whole range by use of a d-c instrument with specially shaped magnet and pole-piece.

The crowding in the lower portion of the scale can be somewhat relieved by the use of a special d-c meter movement. In this movement, the magnetic gap between the permanent magnet and the pole piece is shaped to be greater as the moving coil and pointer swing to higher current values. This makes the instrument progressively less sensitive as current increases, and compensates for some of the low-scale crowding. This is often referred to as "expanding the scale." Figure 7-2 shows how a regular square-law scale can be improved by this process.

7-4. Vacuum Thermocouples

In ordinary thermocouples an appreciable amount of heat is lost to the air surrounding the couple by convection currents (air motion carrying heat). Accordingly, to improve sensitivity, *vacuum* thermocouples are being manufactured. These are mounted inside a glass envelope, resembling that of a small vacuum tube, with both the couple and the heater contained, and leads from each brought out through the envelope. These units are made small enough to be incorporated into the case of some meters or can be connected externally. See Fig. 7-3.

7-5. Shunts and Multipliers for Thermocouple Meters

The range of a thermocouple meter can be increased in the same manner as range increases are made in other types of meters, i.e. by shunting or by series resistance. Shunts and multipliers are used in conjunction with. the heater in such a way that its rating is never exceeded.

It must be remembered, however, that a shunt or multiplier is quite critical, especially if the meter is to be used at relatively high radio frequencies. The least inductance or excess capacitive effect can introduce serious error. Shunts are usually straight, solid wires or bars connected directly across the meter terminals.

Fig. 7-3. An example of vacuum thermocouple used with r-f meters. *Field Electrical Instrument Company*

Thermocouple voltmeters may use several thermocouples in series to provide the required voltage rating. Sensitivities of up to 500 ohms per volt are obtainable.

7-6. Use of Thermocouple Meters on Direct Current

The heating effect in a wire or other conductor is the same for a given resistance and current, whether the current is ac or dc. Thus, although thermocouple meters are most useful for radio-frequency currents, at which there are no other types as suitable, they can still be used on audio-frequency currents and direct current.

However, when a thermocouple meter is used to measure direct current or d-c voltage, a "poling" error may result due to a slight d-c voltage drop being applied to the part of the heater which the thermo-couple contacts. For this reason, the d-c values should be read with the connections one way, then with the meter connections reversed, and the results averaged. This will cancel out the "poling error."

7-7. The Hot-Wire Ammeter

The hot-wire ammeter was described in Chapter 2. This instrument makes use of the warping or sagging of a piece of wire, when it is heated by the electric current it is carrying, to indicate the value of the current. Since it is the heating effect which produces the action, and this applies equally for ac and dc, the meter can be used for either. At one time these meters were widely used as r-f ammeters.

7-8. Use of Diodes for R-F Measurement

Another method for adapting simple meters for r-f measurement is with the use of a diode rectifier, which efficiently rectifies the r-f current or voltage to dc. This dc is then measured with a sensitive moving-coil meter or a vacuum tube voltmeter. Germanium, silicon, and other crystal diodes have low capacitance and are excellent for high radio-frequency rectification. Because these rectifiers are most frequently used with vacuum tube voltmeters, they are discussed in Chapter 9.

8

MULTIRANGE INSTRUMENTS AND VOLT-OHM-MILLIAMMETERS

8-1. Multi-Range Meters

A multi-range meter is a meter movement extended to include two or more ranges. A multi-range meter is distinguished from a volt-ohm-milliammeter (covered later in this chapter) in that it measures only *one* of the basic quantities: current, voltage, or resistance.

The multi-range instrument is in a class by itself in that it is a general utility test instrument. In terms of sensitivity, for example, the d-c voltmeters are in the 1,000 ohms-per-volt class. The alternating-current meters are intended for use at 50-60 cps. The a-c voltmeters are rectifier type instruments in some cases and ordinary moving-vane types in others. Hence the former have a-f applications whereas the latter are intended for 50-60 cps work.

Although these devices find some application in electronic circuits, their sensitivity limits their use. Portability, small size, and low cost are their main features.

Some examples of these instruments appear in Fig. 8-1. Two methods are used for changing from one range to another. One method is to have separate terminals for each range, with one terminal common to all. These are evident in Fig. 8-1A, and will be found on most high current meters, whether a-c or d-c measuring types.

The second method is to use a selector switch. This is commonly done on voltmeters, ohmmeters, and very low current meters. The switch arrangement also is seen in Fig. 8-1B. The difference in range selecting means stems from the effects of resistance in these arrangements. Because devices measuring voltage and very low current are relatively high resistance units, the contact resistance at the switch

Fig. 8-1. Photos of multi-range instruments with terminals and with switch selector.
(A) Triplett, (B) Weston.

terminals is of limited importance. However, in the high current devices it would be important, hence a terminal form of connection is provided.

8-2. Multi-Range Direct-Current Meters

In small portable instruments of the type shown in Fig. 8-1, the meter movements may be as sensitive as 50-microamperes full scale to as high as 200 microamperes, depending on the ranges provided in the device. The multiple functions relative to range are provided by shunts. The switch controlled ranges should not be selected with the current flowing through the meter. The reason for this is that there are usually no shorting switches in a range selector. It is thus possible for the shunt to be momentarily disconnected as the range switch is being manipulated. The circuit current then flows through the basic meter movement coil and, if excessive, can burn out the coil.

Under the circumstances, it is advisable to shut off the flow of current momentarily when switching meter ranges.

It is interesting to note that low-current measuring meters are usable as voltmeters by connecting external multipliers in series with the *lowest* current ranges. In this way the current rating which is used to determine the ohmmic value of the multiplier is the current rating of the range being used. Thus, if the multi-range meter is a micro-

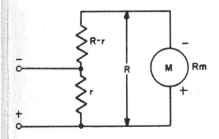

$$r = \frac{R - R_m}{N}$$

r = PART OF R TO BE SHUNTED
     ACROSS MEASURED SOURCE

R = TOTAL SHUNT RESISTANCE

R_m = METER RESISTANCE

N = MULTIPLYING FACTOR

Fig. 8-2. The universal shunt and applicable equation.

ammeter and the lowest current range is 0—50 microamperes, the device can be made into a 20,000 ohms-per-volt voltmeter by using that range. If the instrument is a milliammeter and the lowest current range is 0—1 milliampere, the device can be made into a 1,000 ohms-per-volt voltmeter. A two-gang switch is employed.

8-3. The Universal Shunt

A type of shunt which has come into wide use in conjunction with small, portable, multi-range meters is known as the universal shunt, also called the *ring shunt*, or *Ayrton shunt*. It is readily adaptable for use with meters having different values of internal resistance.

As shown in Fig. 8-2, the resistors in the universal shunt are either connected in series or a single tapped resistor is employed which is shunted across the meter. The resistance of r for any desired factor N, by which the full-scale range of the meter is to be multiplied, is determined by the formula in Fig. 8-2. If the total shunt resistance R is 100 ohms and the internal resistance of the meter R_m is 20 ohms, to increase its current range 10 times the selection of r by means of the top is such that the resistance of r is

$$r = \frac{100 + 20}{10} = 12 \text{ ohms}$$

If an additional range representing 100 times the full-scale range of the unshunted meter is desired, under the same conditions

$$r = \frac{100 + 20}{100} = 1.2 \text{ ohms}$$

The values of the universal shunt shown in Fig. 8-3 are the exact figures required to multiply the range of a 500-microampere meter 2, 20, 200, 2,000, and 10,000 times. Since the meter resistance is 100 ohms, and it is desired to have a 1-milliampere range, it follows that the total shunt resistance must be equal to 100 ohms. When the switch is in the 1-ma position, half the current goes through the total shunt resistance and the other half goes through the meter. The value of each of the shunt resistors is calculated in accordance with the equation in Fig.

8-2. Note that on the 100-ma range, for example, the operative shunt
resistance, r, is equal to the sum of $r3$, $r4$, and $r5$. A similar situation
holds for the other ranges.

8-4. Multi-Range Alternating-Current Meters

The small portable multi-range alternating-current measuring meter
resembles those shown in Fig. 8-1, particularly the one with terminal
connections. The multiple ranges, whether in milliamperes or in amperes,
usually are arranged by means of self-contained current transformers. An

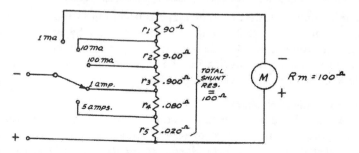

Fig. 8-3. A universal shunt operated by means of a single-gang switch.

example of this appears in Fig. 8-4. This is an inside view of a commercial current meter of the small portable variety. The circuitry of
the system controlling the ranges for one device is symbolized in Fig.
8-5. This is an elaboration of Fig. 6-10.

It is interesting to note that the taps are located in the primary
winding, with terminal A being common to both primary and secondary
circuits. To measure alternating currents up to 1 ampere, the secondary
S acts as a shunt on the meter and the secondary winding is in the current
path. For higher values of current, the free lead is moved to different
primary taps, thus changing the current transformation ratio between
the primary and secondary circuits.

8-5. Multi-Range Voltmeters

Multi-range voltmeters are created by adding series resistance of
the proper value to a basic meter movement. The additional ranges
can be provided by having different *multiplier* resistors available to be
switched into the circuit as desired. A simple arrangement for selecting
multipliers is shown in Fig. 8-6. The switch selects a multiplier resistor
and places it in the circuit, thus providing a voltmeter range.

A more frequently used arrangement is that of Fig. 8-7, in which
all the multiplier resistors are in series across the meter. The external
operation of the switch is the same as in Fig. 8-6. In either case, the
values of the various multiplier resistors are calculated in the same

Fig. 8-4. Internal view of Simpson a-c current meter with current transformer.

manner as explained in Chapter 3. The values of the resistors in Fig. 8-7 are correct for a 1,000 ohms-per-volt, or 1-ma, meter movement whose resistance is 50 ohms. Virtually all small multi-range d-c voltmeters are 1,000 ohms-per-volt.

In the a-c instrument, the arrangement is similar to that of Fig. 8-8. Metallic rectifiers are used and the meter is shunted down by R_s so its effective sensitivity is 1,000 ohms-per-volt (explained in Chapter 3). Thus, for example, between the 1,000-v and the 250-v taps there is a resistance of 750,000 ohms. Since the voltage difference is 750 v, the sensitivity is 1,000 ohms-per-volt.

R_s must be about 10.5 ohms to shunt the 50-μa meter for an effective sensitivity of 1,000 ohms-per-volt. The value of R_v must be such that, added to the value of resistance of the rectifier, the total will be the correct multiplier for 2.5 v, or 2,500 ohms. In practice, R_s and R_v

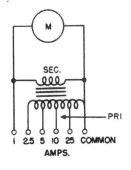

Fig. 8-5. Circuit of multi-range a-c ammeter with tapped current transformer.

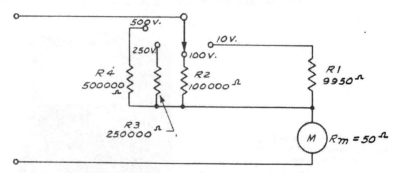

Fig. 8-6. Simple switching system for selecting multi-range voltmeter multipliers.

are wirewound resistors whose exact value is adjusted for the individual rectifier when the meter is calibrated.

Note how the 5,000-v scale "hot" terminal is brought out separately. The selector switch must be on the 1,000-v position for correct 5,000-v readings, since the required 5-megohm resistance can be obtained only by adding all the multiplier resistors in series with the 4-megohm resistor.

8-6. Multi-Range Ohmmeters

Ohmmeters also are available as separate, portable multi-range instruments. In appearance they resemble those shown in Fig. 8-1; the different ranges are switch controlled. Ohmmeter ranges are changed by adding multipliers or shunts to the series-type circuits and also shunts, depending on the individual design. In most instances a single ohms scale calibration is used and the range switch is calibrated in multiplying factors, such as $R \times 1$, $R \times 10$, $R \times 100$ etc.

8-7. Multi-Range D-C Voltage Current and Ohmmeter

This is not a commonplace commercial device, but rather a military unit that warrants comment. The circuit is shown in Fig. 8-9. It is a

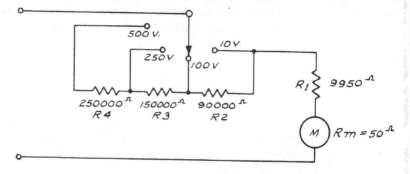

Fig. 8-7. Switching system for series connected voltmeter multipliers, using 50 μa meter for 500-volt maximum scale.

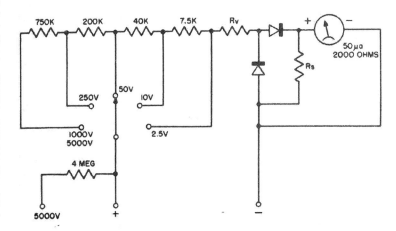

Fig. 8-8. Multiplier arrangement for a multi-range a-c voltmeter.

very simple device with a universal shunt for the a-c current ranges; simple series multipliers for the voltage ranges, and a dual-range series type ohmmeter for resistance measurement. One value of voltage is used for one resistance range and a higher value for the second resistance range.

8-8. Volt-Ohm-Milliammeters (VOM)

These devices are multi-range instruments, but in addition are multi-function too in that they measure alternating and direct current and d-c resistance. Examples of the devices appear in Fig. 8-10.

As can be seen, two physical varieties are available: the small units which resemble the multi-range instruments of Fig. 8-1, and units that

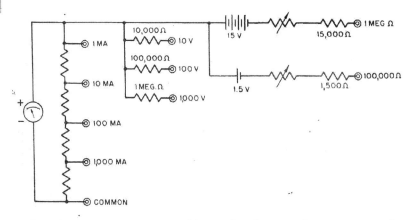

Fig. 8-9. Circuit of combination multi-range d-c voltmeter and current meter used by Armed Forces.

are large physically. Two other differences are found between these groups: the sensitivity of the device relative to its ohms-per-volt rating, and the number of ranges afforded for each function.

Both groups of instruments use d-c moving-coil meters, with metallic rectifiers for a-c measurements. However the smaller and less costly units generally have sensitivity limited to 1,000 ohms-per-volt for both ac and dc, whereas the larger ones afford as high as 20,000 ohms-per-volt sensitivity for dc and up to 5,000 or 10,000 ohms-per-volt for a-c scales.

In all other respects these VOMs are combinations of multi-range meter systems, with one meter being used as the indicator for all functions. Having described the means of adapting meters for d-c and a-c measurements, it is unnecessary to repeat the processes as found in VOMs.

Fig. 8-10. Multi-range and multi-function VOMs: Triplett model 630A and Simpson model 260.

Inasmuch as the chapter on applications of meters deals with the manner of using VOMs, and descriptions of the devices are included, it should suffice here to show circuits of two different brands of VOMs. These appear in Figs. 8-11 and 8-12.

Control of the functions and ranges is by means of a rotary switch, and connection to the instrument is via banana-plug jacks or phone-tip jacks. The selector switch has several decks by means of which the circuit is varied to suit the measurement. Referring for a moment to the two VOMs shown in Figs. 8-11 and 8-12, the d-c voltmeter circuit uses a tapped series multiplier typified by Fig. 8-7. The 5,000-volt range has a separate multiplier resistor which is placed in the circuit when the

CIRCUIT DIAGRAM

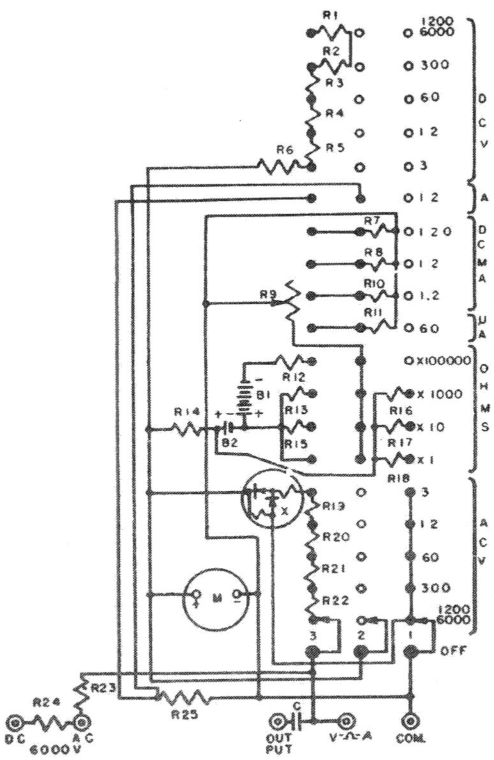

REPLACEABLE PARTS 630-A

Ref. No.		DESCRIPTION	Triplett Part No.
B1	Battery	30V Eveready 413 Burgess U 20 E or Equivalent	T-37-19
B2	Battery	1.5V Eveready 950 or Equivalent	T-2426-1
C	Capacitor	.1 Mfd. 400 DCWV	T-43-69
M	Instrument	50 μa, 250 Mv with panel	T-52-532
R1 } R2 }	Resistor	9 Megohm, ± ½% Carbofilm ½ W	T-15-1541
R 3	Resistor	4.8 Megohm, ± ½% Carbofilm ½ W	T-15-1542
R4	Resistor	960K Ohm, ± ½% Carbofilm ½ W	T-15-1543
R5	Resistor	180K Ohm, ± ½% Carbofilm ½ W	T-15-1544
R6	Resistor	55K Ohm, ± ½% Carbofilm ½ W	T-15 1545
R7	Resistor	2.09 Ohm, Wirewound	T-15-1558
R8	Resistor	20.9 Ohm. Wirewound	T-15-1561
R9	Resistor	20K Ohm Variable	T-16-31
R10	Resistor	217.4 Ohm Wirewound	T-15-1563
R11	Resistor	25K Ohm, ± ½% Carbofilm ½ W	T-15-1546
R12	Resistor	415K Ohm, ± ½% Carbofilm ½ W	T-15-1547
R13	Resistor	380 Ohm, ± ½% Carbofilm ½ W	T-15-1548
R14	Resistor	16K Ohm, ± ½% Carbofilm ½ W	T-15-1549
R15	Resistor	3.16 Ohm Wirewound	T-15-1559
R16	Resistor	5K Ohm, ± ½% Carbofilm ½ W	T-15-1550
R17	Resistor	40 Ohm, Wirewound	T-15-1562
R18	Resistor	4.0 Ohm, Wirewound	T-15-1560
R19	Resistor	45K Ohm, ± ½% Carbofilm ½ W	T-15-1551
R20	Resistor	240K Ohm, ± ½% Carbofilm ½ W	T-15-1552
R21	Resistor	1.2 Megohm, ± ½% Carbofilm ½ W	T-15-1553
R22	Resistor	4.5 Megohm, ± ½% Carbofilm ½ W	T-15-1554
R23	Resistor	24 Megohm, ± 1% Carbofilm 2W	T-15-1226
R24	Resistor	72 Megohm, ± 1% Carbofilm 2W	T-15-1227
R25	Shunt	12 Amp 250 Mv Strip Type	T-90-212
X	Rectifier Assem.	Copper Oxide, Conant BHS-160, 3 Lead	T-2250-13
	Case	Bakelite, with strap handle	T-10-784
	Knob	2 15/32" long, molded (with clip)	T-34-30
	Leads	Banana Plug type with Alligator clips	T-79-49
	Switch	20 Pos. 3 deck, without resistors	T-22-103

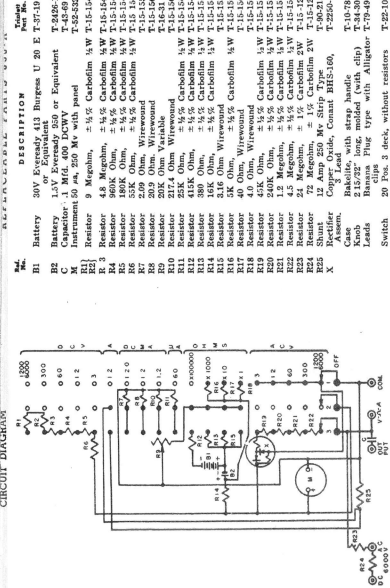

Fig. 8-11. Schematic of Triplett Model 630A VOM.

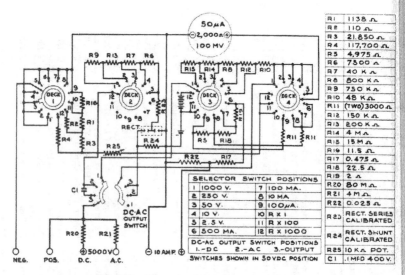

R1	1138 Ω
R2	110 Ω
R3	21.850 Ω
R4	117,700 Ω
R5	4,975 Ω
R6	7300 Ω
R7	40 K Ω
R8	800 K Ω
R9	750 K Ω
R10	48 K Ω
R11	(TWO) 3000 Ω
R12	150 K Ω
R13	200 K Ω
R14	4 M Ω
R15	15 M Ω
R16	11.5 Ω
R17	0.475 Ω
R18	22.5 Ω
R19	2 Ω
R20	80 M Ω
R21	4 M Ω
R22	0.025 Ω
R23	RECT. SERIES CALIBRATED
R24	RECT. SHUNT CALIBRATED
R25	10 K Ω POT.
C1	.1 MFD 400 V.

SELECTOR SWITCH POSITIONS

1	1000 V.	7	100 MA.
2	250 V.	8	10 MA
3	50 V.	9	100 μA.
4	10 V.	10	R X 1
5	2.5 V.	11	R X 100
6	500 MA.	12	R X 1000

DC-AC OUTPUT SWITCH POSITIONS
1.–DC   2.–AC   3.–OUTPUT
SWITCHES SHOWN IN 50 VDC POSITION

Fig. 8-12.   Schematic of Simpson model 260 VOM.

selector switch is in the 1,000 volt position and the test lead is in the 5,000-volt pin jack.

The a-c voltmeter circuit is symbolized by Fig. 8-8; the description that accompanies this figure is a representation of the system used in the device. The ohmmeter circuit is a combination of the shunt and series methods, with different voltage values for the low and high resistance ranges. The direct-current ranges use the universal shunt.

8-9. VOM Scales

A typical VOM instrument scale is shown in Fig. 8-13. Two separate ohms scales are used; the upper one for a series type ohmmeter for high values, and the lower scale is for a shunt type circuit for low values of resistance. As can be seen, these two scales read in opposite directions, for reasons already explained in Chapter 4.

Two scales are provided for voltage measurement, one for a-c volts and the other for d-c volts. However only one series of numbers are used as calibrations. To make it convenient to use the single set of calibration numbers for both a-c and d-c measurements, the corresponding voltage points are joined at the main divisions. The sub-divisions are read in similar manner on both scales.

The basic scale is divided into .2 volt steps with the major divisions being whole number values. The fundamental scale is from 0 to 10 volts. The multipliers increase the range in steps of 5 and 25. The upper scale is read for a-c, and the lower scale is read for d-c, using the same numbers for both.

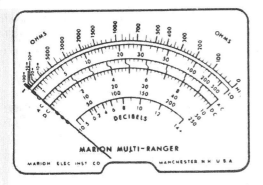

Fig. 8-13. View of Marion multi-range VOM scale.

The direct-current scale is the a-c voltage scale with 10-volts d-c corresponding to 1 ma for one setting of the current range switch and 10 ma for another setting. Similarly, the 50-volt range corresponds to 50-ma and 500-ma full scale respectively. Hence, in one case, the scale calibration is divided by 10; and, in the other case, is multiplied by 10.

Still another scale is used for the db range. By projecting the 0-db point upward to the a-c voltage scale we see that it corresponds to 1.73-volts rms, thus indicating that this scale is for db above or below a reference level of 6 milliwatts into 500 ohms. This is explained more fully in Chapter 6.

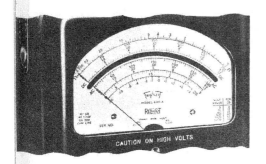

Fig. 8-14. View of scale on Triplett Model 630A VOM.

Another version of a VOM scale is shown in Fig. 8-14. Here only one ohmmeter scale is used and the range selector has calibrated positions for resistance and resistance-multiplying factors. Individual scales are provided for d-c and a-c voltages. In addition a separate 0—3-volt rms a-c scale is furnished for low-voltage measurements. In other brands of VOMs this scale may be 2.5-volts rms. With only three a-c voltage ranges calibrated directly, higher and lower values require the application of multiplying factors or dividing factors as determined by the range selector switch position. Thus to read 600 volts maximum the 60 volt scale is used and multiplied by 10; to read 6,000 volts the same scale is multiplied by 100. This procedure applies to both the d-c and the a-c scales.

8-10. Extending DB Scales

·   The db scale is derived with the meter used as an a-c voltmeter, the only difference being that the db scale is used. Ordinarily, the db scale can be read directly when the selector is set for the lowest a-c voltage range. This reading can always be checked by comparing the 0-db point with the reading on the a-c voltage scale. It is 1.73 volts rms in Figs. 8-10 through 8-14. Frequently this comparison is the only way to determine which voltage scale is to be used for direct db readings.

The same higher scales used for a-c voltage readings are also used for db. For db, however, instead of *multiplying* all direct-scale readings by some power of 10, we *add* the correct number of db. This is in line with the fact that when numbers are *multiplied* the logarithms of the numbers are *added* to obtain the logarithm of the product. Suppose, when reading db, the scale indicates plus 8 db, but the selector switch is set for a voltage scale 100 times the basic scale upon which the db scale is based. 100 times voltage is 40 db (see Chapter 6 for method of calculation). Therefore we must add 40 db to the indication on the db scale with the selector in that position. In this case the reading will be 8 plus 40 or 48 db.

If the original db reading is negative, the *algebraic* sum is used. Suppose the meter range selector is set for a scale 10 times the basic voltage scale and that the indication is minus 5 db. Then, 10 times voltage is 20 db, therefore the indicated level is minus 5 plus 20, or plus 15 db.

9

ELECTRONIC VOLTMETERS AND VOLT-OHMMETERS

One of the most helpful developments in meter type measuring devices for testing electronic equipment is the incorporation of the vacuum-tube in the measuring instrument itself. The basic meter movement (ordinarily the moving-coil permanent-magnet meter) is connected into a vacuum-tube circuit in such a way as to indicate voltage, current, and resistance. The vacuum-tube voltmeter, abbreviated as VTVM, achieves its main advantage over the ordinary meter because it isolates the meter from the circuit under test; and, by utilizing the vacuum tube's capabilities as an amplifier or rectifier, greatly increases the sensitivity of the measuring system beyond that established by the meter itself. Moreover, it extends the frequency range of the meter indicator far up into the hundreds of megacycles region. Finally, it minimizes the loading effect of the measuring device on the circuit under test.

Another advantage of the VTVM is that the input impedance is high and *constant* for all d-c voltage ranges; in contrast to the impedance of the ordinary multi-range voltmeter which varies according to the full-scale voltage range being used.

Most VTVM's have an input resistance of about 11 megohms; laboratory types go much higher. But even in the 11-megohm units this is equivalent to the input resistance of a 20,000 ohms-per-volt meter with a full-scale range of 550 v. However, it is on the *low-voltage* ranges that the advantage of the VTVM is most apparent. For example, a typical 20,000 ohms-per-volt meter has an input resistance of only 0.2 megohms on the 10-v scale; the constant 11-megohm resistance of the VTVM on this scale is 55 times as high. Therefore, we can say that *for low voltage measurements where sensitivity is important, the VTVM is by far the most useful type of meter.* At high voltages, 500 volts dc and higher, the 20,000 ohms-per-volt instrument is just as useful as the

91

vacuum-tube voltmeter. This must be said for the 100,000 ohms-per-volt meter too, which presents a meter resistance higher than the usual VTVM at voltages as low as 150 volts dc.

9-1. The Basic D-C VTVM

The direct-current VTVM is, in the main, an amplifying triode tube used in such a manner that the d-c voltage to be measured causes a change in its plate current. This current is indicated on a moving-coil meter. The voltage to be measured is applied to the control grid circuit of the vacuum tube; it offsets the operating bias and so changes the plate current to above or below the no-input-voltage value. When the plate current meter is suitably calibrated, its indications are interpretable directly in values of applied input d-c voltage.

The schematic in Fig. 9-1A symbolizes the basic arrangement. The conditions of operation are described by the grid voltage—plate current characteristic shown in Fig. 9-1B. Assume that a fixed negative bias just causes plate current cut-off with zero input voltage, the plate current meter indicates zero current. A positive voltage applied to the "input" terminals will offset some (or all) of the bias and cause plate current to flow. According to the characteristic shown, a positive input voltage of 2 volts applied to the grid will increase the plate current from 0 to 4 ma; a positive input voltage of 4 volts increases the plate current to 8

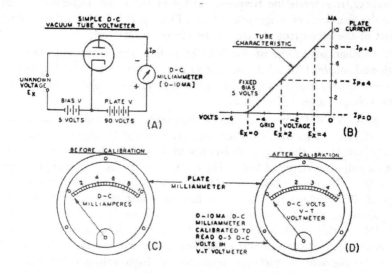

Fig. 9-1. The schematic and operation of a simple d-c, v-t voltmeter. The appearance of the milliammeter scale before and after calibration illustrates the manner in which the plate-current reading is interpreted in terms of unknown voltage applied to the grid.

ma. A maximum input voltage of 5+ volts completely overcomes the operating bias and increases the plate current to 10 ma, the full scale reading of the plate current meter.

The relationship between input voltage and the plate current change enables calibration of the plate current meter in terms of the voltage applied to the input. Thus, instead of calibrating the plate meter in current from 0 to 10 ma, it can be calibrated in voltage from 0 to 5-volts full scale and becomes direct reading. Because the grid voltage-plate current characteristic is substantially linear (within limits normally used) the meter scale is linear.

The limit of d-c voltage which may be applied to the vacuum-tube input circuit is determined by the operating negative grid bias, because freedom from grid current is essential. However, it is readily evident that a meter of this type can neither be restricted in its measuring capabilities to a single range, nor to as low a value as 5 volts. The answers to both limitations are found in the use of a voltage divider across the grid circuit, as shown in Fig. 9-2.

By making the divider resistance very high, say 10 megohms or higher, the input impedance of the instrument is made high. When the taps along the divider are arranged in a suitable manner a wide range of d-c voltages can be measured, and the maximum voltage applied to the grid circuit remains 5 volts for full scale deflection on the meter for every range.

Each tap on the divider affords a certain voltage ratio between the voltage applied across the divider and the voltage applied to the tube. Each tap along the divider represents a voltage range, to be used with its own scale on the meter or as a multiplying factor, as determined by the design of the instrument. In Fig. 9-2 the taps on the divider are arranged in decade steps of 1, 10, 100 and 1,000; these figures serve as multiplying factors. If the lowest full-scale range is 5 volts, the maximum voltage measurable with the device is 5,000 volts when the divider is set to the 1,000 position.

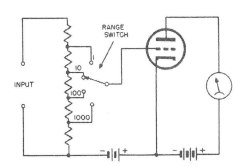

Fig. 9-2. Use of voltage divider in basic circuit.

Referring to Fig. 9-2, it is interesting to note that, if the total divider resistance is 10 megohms, input impedance remains constant for the full range of the instrument, for all scales. Changing the voltage-divider ratio by moving the range switch does not change the total resistance of the divider.

It might be well to comment that subsequent details explain the relationship between the resistance of the divider and the accessory (probe) equipments which are used with VTVMs.

9-2. Center Zero D-C VTVM

The arrangements in Figs. 9-1 and 9-2 require changing the input leads in order to accommodate negative voltages. One method of over-coming the need for changing polarity at the input circuit is the use of the center-zero voltmeter in a fundamental circuit, such as shown in Fig. 9-3. The tube is biased to operate half-way up the grid voltage-plate current curve. The d-c milliammeter (here located in the cathode circuit) accordingly reads in the center of the scale with no external voltage applied to the grid. The point on the meter scale is designated as zero voltage. The meter can now read positive or negative unknown voltages, as indicated on the scale shown in Fig. 9-3B, because the current can increase or decrease as determined by the input voltage.

Control $R9$ adjusts the plate voltage so that the meter reads zero with no external voltage. The grid is self-biased by cathode resistors $R6$ and $R7$. $R6$ helps stabilize the circuit by providing a suitable amount of degeneration, and control $R7$ is used to compensate for minor circuit variations. Selector switch $S1$ and a voltage-divider network enable the voltmeter to operate over four different voltage ranges, and an a-c filter eliminates any a-c component in E_x.

The center-zero arrangement has the advantages of simple design and low cost, plus elimination of the need for a polarity switch to measure both positive and negative voltages.

The circuitry used in commercial d-c VTVMs differs from these basic arrangements, even if the fundamental operating techniques do

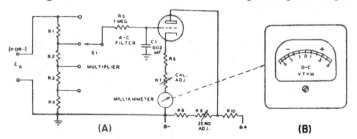

Fig. 9-3.   (A) A center-zero instrument. (B) Typical scale calibration.

not. For instance, the modern day commercial instruments make use of two or more tubes in the metering circuit. In modern instruments these are bridge circuits, which are explained briefly in this chapter. The reason for the more elaborate circuits is to develop the required stability in operation, especially the zero plate current settings, thus assuming maximum accuracy and achieving utmost sensitivity.

The vast majority of these instruments are a-c operated. Line voltage variations, differences in tube constants, changes in circuit elements, etc., will affect the zero setting, and the performance of the device, adversely unless steps are taken to use such circuits as minimize these effects.

9-3. The Two Tube Bridge Circuit

A circuit commonly used in the d-c VTVM is shown in Fig. 9-4A. The vacuum-tube system uses two tubes in a bridge circuit. Two arms of the bridge are the cathode resistors of the two tubes. The other two arms of the bridge contain the vacuum tubes themselves; one of these is the active tube, and the other tube, of identical characteristics, is used as a balancing tube in the other arm. The meter is placed between the cathodes, that is, between the tube ends of the series-connected cathode resistors.

The resistor R3 makes it possible to compensate for variations in the tubes. With no voltage input the plate current indication is zero, because there is the same difference of potential across each of the two cathode resistors. The plate resistances of the two tubes are alike, hence the entire circuit is in balance. When a voltage is applied to the input of the active tube, its plate resistance is changed and the bridge balance is upset. Current then flows through the meter because a difference of potential exists across the tube ends of cathode resistors. The resultant current flow through the meter can be interpreted in voltage input applied to the active tube. This is done by suitably calibrating the indicating meter in input volts.

Although the circuit in Fig. 9-4A does not show it, a voltage divider is found in the input circuit of the active tube. The resistor R4 is the series-multiplier resistor which makes the current meter suitable for use as a voltage meter; and, if made variable, affords a multiplicity of voltage ranges for the meter.

A variation of the two bridge circuit is shown in Fig. 9-4B. In this case the indicating meter is connected across the tube ends of the two plate load resistors R1 and R2. The potentiometer R3 serves the same purpose as before, to compensate for variations in the two tubes. The resistors R4 and R5 are the two cathode bias resistors; R6 is the relatively high value common cathode resistor. These three cathode resistors contribute to the degeneration in the system.

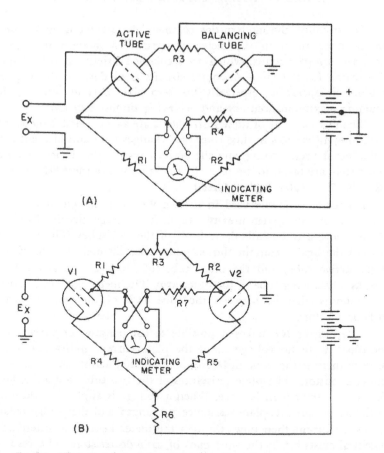

Fig. 9-4. (A) A two-tube v-t voltmeter in which the resistance arms are connected in series with the cathodes providing degeneration as well as a bridge balance. The meter is connected between the cathodes. (B) Two-tube push-pull bridge d-c amplifier, with meter connected between the plates.

With no voltage input, the meter current is zero because no difference of potential exists between the ends of *R1* and *R2* where the meter is connected. The voltage to be measured is applied to *V1* (the active tube), causing its cathode current and effective plate voltage (hence plate resistance) to change. The changed cathode current of *V1* flows through the common cathode resistor *R6*, and causes the plate voltage of *V2* to change in the opposite direction to the plate voltage change in *V1*. Hence the difference of potential across the terminals of the indicating current meter is twice (once for *V1*, and once for *V2*) what it would be if only one tube alone were used. This is the result of push-pull action.

The circuit is not sensitive to minor changes in tube electrode voltages stemming from the operating voltage sources, hence the zero indication is stable to a satisfactory degree.

Although described as d-c VTVMs, virtually all of these devices can measure d-c resistance and a-c voltage. As is shown elsewhere in this chapter, the a-c measuring VTVM uses the d-c amplifier system as part of the overall operating circuit — to measure the a-c voltage after it has been rectified.

9-4. Direct-Reading V-T Ohmmeter

By addition of the circuit shown in the 'left-hand portion of Fig. 9-5A almost any type of d-c VTVM is converted into an ohmmeter. The VTVM shown in the right-hand portion of the figure is a symbolized balanced circuit. It can just as readily be the bridge circuit of Fig. 9-4. Before R_x, the unknown resistance, is placed in the circuit the full-

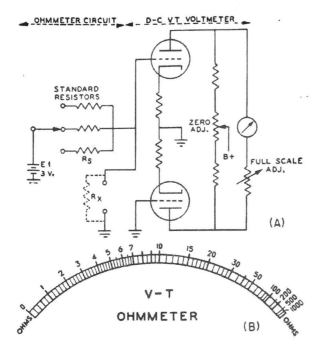

Fig. 9-5 (A) A direct reading v-t ohmmeter circuit. The unknown resistance is determined by measuring the voltage developed across it. A typical scale is shown in Fig. 8-9. Fig. 9-5 (B) Typical scale of a vacuum-tube ohmmeter using the basic circuit shown in Fig. 8-8. This scale is used in the instrument shown in Fig. 8-7 to cover the range from 0.1 ohm to 1,000 megohms.

scale adjustment is made. In this condition the full bias voltage of $E1$ is applied to the grid of the upper triode through one of the standard resistors, R_s. If R_x is now inserted in the circuit, R_s and R_x form a voltage divider across $E1$, and only the voltage drop across R_x is applied to the upper triode. The meter will then indicate something less than full scale, the amount being directly proportional to the decreased bias on the grid of the upper triode, and of course directly proportional to the resistance of R_x. The scale can then be calibrated in terms of the resistance of R_x. A typical scale calibrated in this manner is shown in Fig. 9-5B. Remember that before a reading is taken *this ohmmeter is adjusted for full-scale* without R_x in the circuit, thus signifying an infinite resistance.

9-5. The A-C VTVM

The VTVM used for measurement of a-c voltages involves a number of considerations in addition to those encountered in the d-c measuring device. Finished instruments allow the measurement of both a-c and d-c quantities, but the former entails rectification, whereas the latter does not. Thus, while a single VTVM may provide these two measuring capabilities, the a-c voltage measuring facility stems from a somewhat more complicated circuit arrangement.

The voltage of a rectified wave is proportional to the rms voltage of the a-c wave from which it is derived, hence magnitude is determinable after rectification. Whether amplification, which increases sensitivity, precedes or follows the rectifying action, is a function of design.

9-6. Series Diode Circuits

The simplest a-c VTVM circuit is that using a diode. The diode, of course, provides no amplification, its primary application being that of *rectification*. After conversion to a pulsating d-c voltage this a-c voltage causes the d-c meter to function and, by suitable calibration, to indicate the input a-c voltage.

The series type diode VTVM has the same circuit as the copper-oxide rectifier a-c meter discussed in Chapter 6. Like the copper-oxide rectifier, the diode characteristic tends to be *non-linear,* at *low values* of applied voltage and with a low value of load impedance. Figure 9-6 shows the diode characteristic with no load resistance (curve *a*), with medium load (curve *b*), and with high load (curve *c*). Curves *b* and *c* show that non-linearity may be almost eliminated by addition of a series resistance R of sufficiently high value. Unfortunately this lowers the overall sensitivity of the circuit, but is still a practical instrument.

Figure 9-7 shows a series type diode circuit and the mode of operation on the diode plate voltage-plate current characteristic. The load resistor R has been made high enough to insure linear response, as for

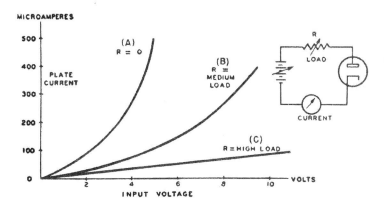

Fig. 9-6. Plate current—plate voltage characteristic of a diode under different loads.

curve *c* in Fig. 9-6. The meter movement responds to the average rectified current, as explained in Chapter 6.

The sensitivity of this type of circuit is limited by the series resistor, *R*. Its frequency response is reasonably good up to the lower radio frequencies, where capacitive effects begin to interfere. At the higher frequencies the meter tends to follow *peak* voltages, due to stray capacitance. By adding a suitable capacitor across the diode load resistor (Fig. 9-8), the series-type circuit is made to respond to the *peak* value instead of to the average value. Note that the circuit then is practically the same as the detector of an a-m receiver. *C* must have sufficient capacitance to make its reactance low compared to the resistance of *R* at the frequency of the unknown voltage. The effect of this circuit on the rectified waveform is shown in Fig. 9-9.

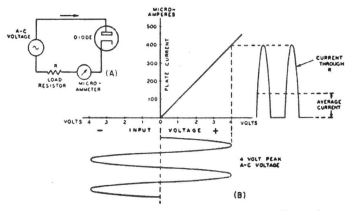

Fig. 9-7. The operation of a simple series type diode rectifier circuit.

9-7. Shunt-Type Diode Circuits

When a-c voltages are to be measured in circuits which also contain direct current, the dc must be kept out of the instrument. In the series-type circuit just described there is a d-c return path between the measuring terminals, although d-c can be kept out of the voltmeter by means of a suitable series capacitor in the "hot" lead. Direct current can be kept out by use of the shunt-type circuit of Fig. 9-10. The a-c voltage to be measured is coupled through a suitable d-c blocking capacitor C to the plate of the diode, causing the tube to pass current on positive half-cycles. The tube conducts through the only available d-c path, that is, through the load resistor R and the meter M. The capacitor C keeps direct current from the voltage source from affecting the readings. This arrangement inherently responds to *peak* a-c voltage, because the charge time of the capacitor is longer than the time for a cycle of the a-c voltage.

9-8. The Triode-Type VTVM

In the diode-type circuits considered so far the lack of amplification limits the sensitivity of the instrument. Furthermore, because they must draw current from the circuit under test, the diode-type circuits have a definite loading effect.

Both of these limitations are overcome by the triode-type VTVM circuit. The amplifying ability of the triode circuit is used to provide greater overall sensitivity. Also, by operating the grid so that it draws no current, a high-impedance input circuit is provided. The voltage to be measured is applied between the grid and cathode, and (except for cathode followers) the meter responds to the resulting plate current flow.

Three different modes of operation are possible, depending on the way the triode is biased, as illustrated in Fig. 9-11. When the operating point is at *A* the operation is called *half-wave*, because the negative half-cycles are cut-off just as in the diode-type circuit. In fact, except for the

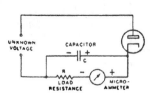

Fig. 9-8. A series type peak VTVM.

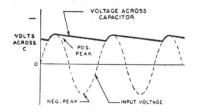

Fig. 9-9. Effect of shunt capacitor C on rectified waveform.

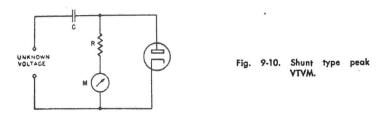

Fig. 9-10. Shunt type peak VTVM.

amplification provided, the operation is equivalent to that of a diode with a relatively small load resistance. With bias at *B full-wave* operation is achieved, because there is plate-current flow on both the positive and negative half-cycles of the voltage to be measured. For so-called "peak operation" at *C*, plate current flows only on the tips of the positive peaks of the a-c voltage applied.

9-9. Half-Wave Square-Law VTVM

The basic circuit and operation of the half-wave square-law VTVM are shown in Fig. 9-12. As the designation *half-wave* implies, the operating grid bias is chosen close to cutoff so that the plate current is essentially zero when no signal is applied to the grid. When a signal is applied the plate current increases during the positive half of the cycle, as is shown at (*b*). On the negative half. however. no appreciable decrease in the plate current can take place. As a result of this rectification, the average value of the plate current increases when a signal is applied, and this increase can be taken as a measure of the voltage applied to the grid.

The outstanding characteristic of this type of VTVM is that the plate current is approximately proportional to the *square* of the voltage applied to the grid. This makes the calibration independent of the waveform, although there may still be some waveform error due to the phase of the harmonics present in the unknown voltage. Compensation

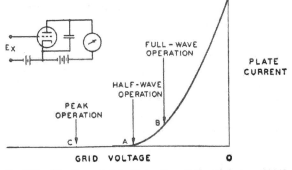

Fig. 9-11. The three modes of operation of the triode type VTVM.

for variations in waveform cannot be complete in the half-wave type of
VTVM because the waveform of the negative half-cycles cannot directly affect the reading of the voltmeter.

9-10. Rectifier-Amplifier Type

Fundamentally all a-c VTVM measurements depend · upon rectification of the a-c voltage. As mentioned earlier, the advantage of rectifying the a-c voltage is that it permits the use of less expensive d-c
instruments. At the same time, the conversion of the unknown a-c
voltage into a d-c voltage has the important advantage of enabling
measurements to be made over a very wide range of frequency.

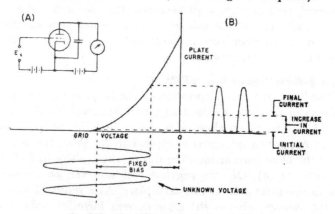

Fig. 9-12. The half-wave square-law VTVM.

In the VTVM's described previously, a basic rectifier stage is used
with the d-c meter inserted so as to measure the rectified plate current.
By adding a d-c amplifier stage, as shown in Fig. 9-13, however, the performance can be greatly improved. The addition of the d-c amplifier has
the advantage of providing higher sensitivity, greater stability, a wider
range, and, in certain instances, higher input impedance. These advantages can be realized because the separation of the functions of rectification and amplification make it possible to obtain the greatest efficiency in each.

The rectifier-amplifier type of VTVM is particularly well adapted
for a probe type of input system. The probe is the movable extension of
the measuring device, and contains the rectifying device. By joining the
probe to the meter system through a flexible cable, ready access to the
circuits to be measured is made available. Moreover, by bringing the
rectifying device to the circuit under test, short leads between the circuit
and the measuring device are possible, and very wide range of fre-·
quencies can be covered.

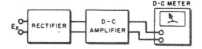

Such rectifying means in probes are either vacuum tube diodes or germanium crystals. Since the leads from the rectifier to the metering circuit carry only d-c (and 60 cps a-c for heaters of vacuum tube diodes), this connecting cable is not critical in length. The illustrations in Fig. 9-14 show a vacuum tube (A) and a crystal (B) as used in a-c measuring VTVM probes.

9-11. Probes for D-C and R-F

A probe is a device which can be added to the input of an electronic voltmeter to facilitate the measurement of radio, intermediate, VHF and higher frequencies, and d-c high voltages. There are various types of probes, each designed for a specific purpose. In general, probes are designed either to increase the input resistance or to decrease the input capacitance of the voltmeter, hence to minimize circuit loading. In particular, d-c isolating probes minimize the effect of d-c measuring-equipment leads in circuits carrying a-c or r-f signals.

Figure 9-15 shows the circuit of a simple probe consisting of a resistor in series with the cable input lead of the voltmeter. The resistance most commonly used for this purpose is 1 megohm. If the d-c input resistance of the voltmeter is 10 megohms, 10/11's of the voltage applied to the probe tip will appear across the voltmeter terminals, and the voltmeter calibrated to read correctly under these conditions. A direct connection to a voltmeter so calibrated yields a reading 10 percent high.

The use of this resistor permits the probing of circuits without loading or detuning them. This is particularly important for the measurement of d-c bias in the grid circuit of an oscillator where loading is undesirable. A resistance of 1 to 5 megohms is enough to prevent resistively loading most circuits; the shunt capacitance of the resistor, usually of the order of 1 $\mu\mu$f, does not greatly detune or otherwise affect most circuits. The resistor also increases safety in making measure-

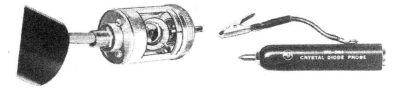

Fig. 9-14. Diode probes: (A) vacuum type; (B) crystal type.

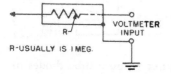

R—USUALLY IS 1 MEG.

Fig. 9-15. Schematic diagram of a simple d-c probe.

ments, since it will effectively limit the current in case of bodily contact with lead or voltmeter terminals.

The range of a voltmeter may be extended to very high voltages by increasing the value of the probe resistance. A probe resistance of 991 megohms, for example, will multiply the range of a voltmeter having an input resistance of 10 megohms by a factor of 100, so that a 100-volt full-range scale will read up to 10,000 volts. (A 991-megohm resistor is mentioned because most meters are calibrated for use with a probe resistor of 1 megohm; this value must be added to the multiplier used.) The length of the probe must be increased so that it will accommodate a long 991-megohm resistor.

The longer the resistor and probe, the less chance is there that the probe will break down or arc over with 10,000 volts across it. A barrier is usually provided on these probes to prevent the user from handling the tip and making contact with the high voltage being measured (see Fig. 9-16). These probes drop the voltage to a safe point at the VTVM terminals and limit the current to a safe value. For instance, a 1,000-megohm probe resistor limits the current from a 30,000-volt source to 30 microamperes. An additional precaution used in some probes is the inclusion of a grounded shield in the handle extending up inside the barrier. In case of an arc-over the spark will jump to the grounded shield rather than to the user's hand.

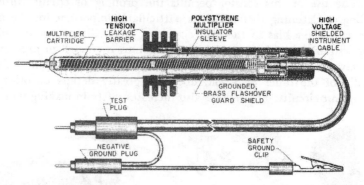

Precision Apparatus Co.

Fig. 9-16. Cross section of a multiplier type d-c probe, showing the long ribbon-wound resistor, the guard shield, and the barrier. Note that the resistor extends almost the the full length of the probe.

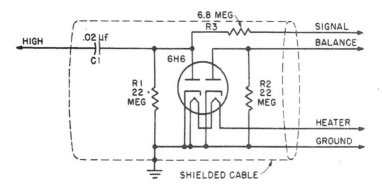

Fig. 9-17. Circuit diagram of diode-rectifier probe used for a-c and r-f measurements.

9-12. A-C Probes and R-F Probes

Figure 9-17 is the circuit diagram of a typical diode-rectifier probe used for a-c and r-f measurements. The input impedance is derived from the combination of shunt and series resistors and capacitors shown in the signal-diode circuit. The second diode is used to balance the contact voltage of the first.

Suitable input terminals, one of which is usually grounded, are provided on the probe; jacks, clip leads, or short wires soldered to the circuit under test, are used for connecting purposes.

Since a probe permits the use of very short connecting leads, its widest application is with high-frequency voltmeters. Since early VTVM's were single tube affairs in which rectification and d-c amplification or meter coupling were accomplished in a single tube, and since early tubes were large and awkward, the first probes were cumbersome. Gradual development has evolved smaller and smaller probes mounted on more and more flexible cables.

Figure 9-18 shows a well designed probe capable of measuring r-f voltages to about 700 mc. The tube used in this probe was specially designed to reduce stray capacitance, stray inductance, and other error-introducing factors. At frequencies above a few hundred kilocycles, long

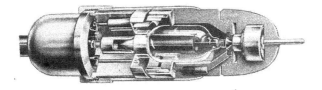

Fig. 9-18.   Cross-sectional view of an r-f and high-frequency probe used for measurements up to 700 mc. Hewlett-Packard

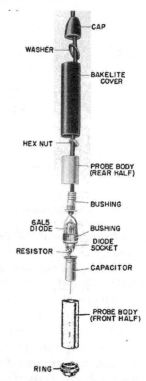

Fig. 9-19. Disassembled RCA duo-diode probe used for measurements from audio frequencies up to about 250 mc.

connecting leads introduce such errors and increase the circuit loading. Therefore the probe shown in Fig. 9-18 is not only interesting from a design standpoint, but also is necessary for maximum accuracy at high frequencies.

The probe shown in the cutaway view in Fig. 9-19 is the RCA model WG-275. Audio-frequency voltages up to 100 volts rms can be measured with this unit. Its upper frequency limit is above 250 mc.

9-13. Crystal Probes

A convenient type of rectifier found in probes used with VTVMs is the germanium crystal "crystal probe." A typical circuit is shown in Fig. 9-20. It is intended for peak voltage measurements on sine-waveform voltages. A peak-to-peak measuring probe is shown schematically in Fig. 9-21.

Although crystal devices are used at frequencies in the microwave region, the general run of crystal probes are intended for frequencies up to about 250 mc. Unlike the tube diode, the crystal probe requires no heater excitation, hence the probe carries only the rectified (pulsating d-c) signal, and the lead length is not too critical.

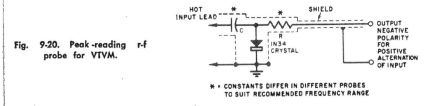

Fig. 9-20. Peak-reading r-f probe for VTVM.

* = CONSTANTS DIFFER IN DIFFERENT PROBES
TO SUIT RECOMMENDED FREQUENCY RANGE

Crystal probes recommended for VTVM use are basically peak rectifiers. The lowest frequency of operation is determined by the time constant of the probe circuit itself. The frequency expressed as time interval $(t = 1/f)$ should be at least 1/10 of the time constant of the probe to afford peak rectification.

When only the modulation component of a modulated r-f carrier is to be measured, it is necessary that the crystal probe meet the following requirements: The time constant of the probe has to be at least equal to 10 times the modulation frequency; also the probe time constant should be 1/10 or less of the carrier frequency. Only when these two requirements are met will the probe give the desired results.

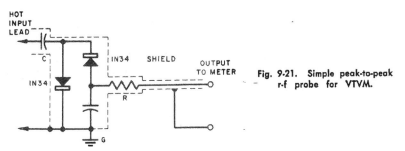

Fig. 9-21. Simple peak-to-peak r-f probe for VTVM.

The modulation and carrier frequencies have to have a ratio of 1:100 or better to permit peak detection. Whenever the time constant of a crystal probe is calculated, the shunt capacity of the leads and the input impedance of the meter (input resistance and capacity) have to be taken under consideration. A crystal probe should be considered as a variable load on the circuit where it is used for measurement. The actual impedance of the probe varies with frequency. That is mainly due to the reactance of the blocking capacitor C.

The maximum voltage which can be applied to crystal probes is again determined by the reactance of the blocking capacitor at the frequency of operation and the inverse peak voltage rating of the crystal. It is obvious that the blocking capacitor and the crystal act as a voltage divider which has a variable ratio depending upon the frequency of operation. In no case should the peak voltage appearing across the

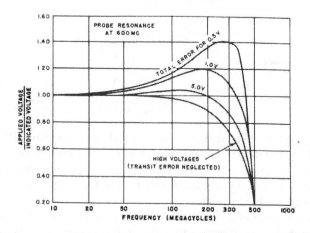

Fig. 9-22. Graph of voltmeter reading error versus frequency of measured voltage for various voltage values measured on a VTVM using a diode rectifier probe with a 5647 tube. *Sylvania*

crystal be greater than the maximum peak inverse voltage of the crystal used in the probe.

Usually a crystal probe for a VTVM contains a series resistor in the probe-to-VTVM lead. The function of this resistor is twofold. First, this resistor provides isolation between the probe and the meter and, secondly, permits adjustment so that the meter reading is correlated to the voltage applied across the probe.

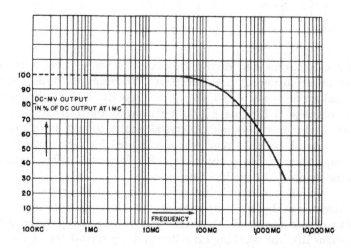

Fig. 9-23. Frequency correction curve for the Millivac Instruments model MV-18B using crystal probes.

Just what this voltage is for different crystal probes is stated by the manufacturer. As a generalization, safe maximums should not exceed 50 volts, although some crystal probes have been rated as high as 100 volts.

9-14. Frequency Correction Curves

While the useful frequency range of VTVM diode probe heads is broad, the accuracy of indications over the frequency range of these instruments is not uniform. Accordingly, some test equipment manufacturers provide frequency correction curves that correlate frequency and voltage reading. In this way, the actual voltage present at the probe tip can be determined. Examples of these correction curves appear in Fig. 9-22 and Fig. 9-23. Complete data of this kind accompany instruments which require them.

9-15. Input-Resistance Correction Curves

The input resistance of probe heads does not remain constant over a wide band of frequencies; hence it is important to know what

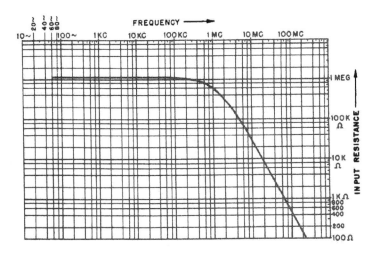

Fig. 9-24. Input resistance vs frequency curve for RCA type WV-75A.

correction factors must be applied. This information is furnished with instruments which are designed to operate over a frequency band in which the change in input resistance is substantial. An example of an input-resistance correction curve appears in Fig. 9-24. Note that the input resistance remains constant over a range from about 60 cps to 500 kc, then begins to fall. At 100 mc the input resistance of the probe head falls to 700 ohms. This change in input resistance must be taken into account if correct measurements are to be obtained.[1]

9-16. Summary of Electronic (or VTVM) Meter Types by Use

Because of the great versatility inherent in the electronic meter (whether it is a vacuum-tube voltmeter or a transistor voltmeter, or other electronic type), there has evolved an extremely wide variety of instrument forms using this type of electronic circuitry, all grouped under the general heading VTVM. Since many of the electronic instruments are designed to provide special characteristics, and others are especially suitable for advanced laboratory work rather than for run-of-the-mill technical work, it is well to separate the discussion into three meter groups,[2] as follows:

Group I, the Laboratory Group. This covers a wide variety of instruments desirable for laboratory use without being too specialized in their voltage-measuring function. Models in this group incorporate various types of multistage amplifiers and generally fall in a price class above $150.

Group II, the Service Group. These meters usually employ a dual triode for a balanced bridge and are widely used for general-purpose maintenance measurements. Instruments in this group are usually priced at less than $150.

Group III, the Special Group. This catch-all designation includes the more specialized types. These would undoubtedly find use in either service or laboratory work, but they are separated from the first two groups by their special features.

9-17. Summary of VTVM Circuit Types

The circuit used for an electronic meter in any particular group generally follows a discernible pattern, so that it will be profitable to consider the electronic circuit for a given instrument as falling into a basic circuit category, independent of its instrument group. Although there are four of these basic circuit types to be distinguished, the picture is not as complicated as it first appears, since the majority of the instruments in each group generally favor one of the circuit types over the others, as will be pointed out later. The four basic circuit types are as follows (see Fig. 9-25):

---

[1] For a more detailed and thorough discussion of probes in testing, see A. Ghirardi and R. Middleton, "How to Use Test Probes," John F. Rider, Publisher, Inc., New York, 1955.

[2] For an extensive listing of commercial instruments arranged according to these groups, refer to S. D. Prensky, "VTVM Survey," *Electronic Design Magazine*, Sept. 15, Aug. 1, and Aug. 15, 1957 issues.

Circuit Type 1. This circuit is a *straight d-c amplifier* and is generally used in a symmetrically balanced form. Amplifier-gain stability is greatly improved by feedback, especially when more than one stage of amplification is used. Grid current, which is an inherent characteristic of the input tube, is not changed essentially by a feedback, as is sometimes believed. In spite of grid-current limitations of around 10^{-11} amp, many special applications using electrometer type tubes are in commercial use. Details of this advanced type of VTVM are discussed in Chap. 12, under high-sensitivity d-c measurements.[3] (When the single-stage balanced-bridge is preceded by a rectifier, it becomes a type 2 circuit, as explained next).

Circuit Type 2. Consisting of a *rectifier plus a balanced-bridge d-c amplifier* this is the same circuit described in the previous sections. It is widely used where a combination of a-c and d-c volt ranges is desired. It can be readily adapted to resistance and peak-to-peak ranges, especially for service work. When used for laboratory a-c measurements, this circuit is capable of covering a frequency range up to 300 mc, if maximum voltage sensitivity is not a prime consideration. The non-linearity of the rectifier limits accuracy in low-voltage measurements. Lowest-full scale range is usually 0 to 1 v, if the wide-frequency capability is retained.[4]

Circuit Type 3. Particularly suitable for measuring small a-c voltages, this type employs an *a-c amplifier-plus-rectifier* circuit. It has wide use in audio- and low-frequency applications, where it is often provided with logarithmic meter-indications. For laboratory use, it appears most often in an arrangement having a frequency-compensated, constant-impedance input for a multistage a-c amplifier, with large amounts of feedback to stabilize gain and expand band width. Shunt capacitance in the input circuit is fairly high (around 25 $\mu\mu f$), but generally acceptable for applications up to a few megacycles. Lowest full-scale ranges are usually around 1 to 10 mv. Overload protection for the meter is obtained from the buffer action of the amplifier. This type of advanced instrument is described in Chap. 12.[5]

Circuit Type 4. This type uses *a modulator* for conversion of dc to ac followed by a-c amplification, *plus a rectifier circuit.* It is capable of very high d-c voltage sensitivity at a high input impedance. With chopper d-c to a-c conversion, only a narrow frequency response of the a-c amplifier is required, since it need cover only the driver frequency, as expanded by the rate of the d-c input variation. Use of large amounts of feedback in the multistage a-c amplifier allows improvement of sta-

[3] *Ibid.*
[4] *Ibid.*
[5] *Ibid.*

bilized gain and higher input impedance. In a pH application, input impedance can be made to exceed 100 meg.[6]

9-18. Measuring Video-Frequency Signals

When measuring signals in the video-frequency (v-f) band, extra attention should be paid to the frequency range of the electronic meter being used, because of the different capabilities of instruments in this range. Although the expression *video-frequency signals* is quite definite

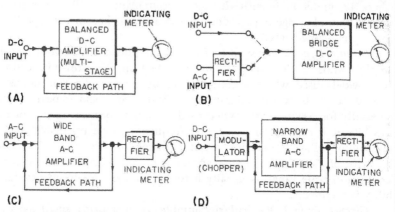

Fig. 9-25. Main VTVM circuit types: (A) multistage d-c amplifier; (B) rectifier plus d-c amplifier; (C) multistage a-c amplifier plus rectifier; (D) d-c modulated type.

when applied to actual video signals in a television set, covering a range from d-c out to around 4 or 5 mc, the expression is also widely used in a loose sense for those signals between the upper end of the a-f range (around 20 kc) and the lower end of the r-f range (around 0.5 mc). The term *intermediate frequency* (i-f) is also used in a loose sense in these frequency areas, but it is even less useful in defining a definite range within which a meter should be used, because it might designate a frequency of 175 kc for an a-m radio set, 10.7 mc for an f-m, or even one as high as 45 mc for a tv receiver.

For the purposes of this discussion of meter types, the term *video frequency* can be taken to apply to those frequencies that go beyond the strictly audible range of 20 kc and up to the video-signal limit of around 4 or 5 mc. The need for caution in selecting a meter for this range arises from the fact that the ordinary general-purpose service-type of VTVM, when used without a special probe, may have a frequency limit as high as 4 mc or as low as 0.5 mc, depending upon its

---

[6] *Ibid.*

make. Wherever possible, it is advisable to consult the manufacturer's instruction sheet to determine the instrument's upper frequency limit (and its impedance at that range), to determine whether it is capable of an accuratae measurement at the frequency desired. When in doubt, it would be safer to use an accessory r-f probe for any measurements at frequencies higher than 500 kc. Of course, if only relative deflections are desired ,the instrument may be used without the accessory probe, for comparative measurements, all made at the same frequency, as long as a reasonable amount of deflection is obtained at the frequency in question.

9-19. Pulse and Complex-Wave Indications

In addition to the frequency consideration in a-c measurements, the waveform of any signal that is not a pure sine wave must be considered. The effect of waveform on meter indication in terms of the 1.1 form-factor (which gives the ratio of the rms value of a sine wave to its average value) was discussed briefly in Sec. 6-9 in connection with the moving-coil rectifier-type meters. The situation is different however when dealing with the general-purpose type of VTVM where the rectifying diode is most usually arranged in a circuit that produces a d-c output voltage proportional to the *peak amplitude* of the input sine wave. After amplification, this d-c value is displayed on the meter scale, which is generally calibrated in terms of the rms value of a sine wave, equivalent to this peak voltage.

With a pure sine-wave input signal, this circuit provides a highly satisfactory degree of linearity between the rms meter indication and the rectified peak amplitude. However, the meter indicator is very sensitive to the presence of harmonics in the applied wave. For example, the addition of a 20% harmonic to a pure sine wave can increase the meter indication by a much larger percentage than is warranted by the increase in rms voltage, such as would be indicated by a square-law thermocouple meter. Also, the phase relation of the harmonic to the fundamental determines whether the reading will increase or decrease. It is thus possible for a 20% second harmonic to increase the reading by as much as 20% or decrease it by as much as 10%, while a full-wave square-law thermocouple meter would show only a modest increase of 2% regardless of the phase of the harmonic.[7] It is therefore necessary to use extra care in interpreting the meter reading whenever complex waveforms are being measured, unless only relative readings are needed. For reading the absolute value of a complex waveform, most electronic

---

[7] For a comparison of indications on various types of voltmeters for second and third harmonics, refer to F. E. Terman and J. M. Pettit, "Electronic Measurements," 2d ed., Table 1-1, McGraw-Hill, New York, 1952.

voltmeters provide a *peak-to-peak function*, which provides a satisfactory reading of the over-all amplitude excursion of the complex wave. Note carefully, however, that this p-p reading must be supplied as a separate switch position on the function switch. If a meter dial carries a p-p scale that is simply a repetition of the rms scale multipled by a scale factor of 2.8, but has no separate peak-to-peak position on its function switch, it is no help at all in obtaining the true absolute p-p value of a complex wave. If the approximate percentage and order of the harmonics present in the input signal are known, the amount by which the meter indication will deviate from the true value can be determined from a table, usually given in the instruction manual. Data for a laboratory-type VTVM employing a full-wave rectifier arrangement are given in Table 9-1.

TABLE 9-1.   EFFECT OF HARMONICS ON VTVM INDICATION*

Input voltage characteristics	True rms value	Value indicated by model 400D VTVM
Fundamental = 100	100	100
Fundamental + 10% second harmonic	100.5	100
Fundamental + 20% second harmonic	102	100–102
Fundamental + 50% second harmonic	112	100–110
Fundamental + 10% third harmonic	100.5	96–104
Fundamenatl + 20% third harmonic	102	94–108
Fundamental + 50% third harmonic	112	90–116

* Courtesy of Hewlett-Packard Company.

10

TYPICAL APPLICATIONS OF METER FUNCTIONS

In this chapter we review the most important ways in which meters can be used, with emphasis on *electronic* circuits. First, however, let us compare the general features of each of the three main types of meters: the panel-type meter, the volt-ohm-milliammeter (VOM) and the VTVM so there will be no doubt as to which is most appropriate for each application.

10-1. Panel Instruments

Typical panel-type meters are illustrated in Fig. 10-1. The three types are designed to be mounted permanently in the equipment in which they are used. In most cases the meter movements are designed to operate best with the meter face in a vertical plane. A mechanical zero adjustment is provided as shown in the figure. This is used to adjust the pointer to a zero scale reading when no current is flowing. It should be adjusted with the meter positioned in the plane in which it will be used.

Parallax. The refractive quality of the glass face cover may lead to an effect called *parallax,* which makes the pointer appear to be in two places at the same time. Because this can lead to error, the meter should be read from *directly in front,* or directly down. Looking at the meter face at an angle will result in erroneous readings. Parallax is minimized in the more expensive varieties by using a knife-edge pointer riding across a strip of mirror which reflects the pointer image. When the meter is viewed properly the pointer is in line with its image in the mirror.

External Magnetic Fields. Meters are magnetic instruments and can be influenced by external magnetic fields. Therefore a meter should not be mounted close to current-carrying filter chokes, power trans-

Fig. 10-1. Photographs of typical, modern panel meters. (Note: The above photos are reproduced through the courtesy of the manufacturers indicated.)

formers, or even near large pieces of iron or steel. The meter panel itself, if steel, can influence the readings and must be taken into account. For this reason most meter manufacturers state the type panel on which the meter may be mounted, and whether steel panel mounting is recommended *at all*.

Potential to Ground. A factor in whether a meter is to be mounted on *any* kind of *metal* panel is the potential to be applied to the meter circuit. For example, if current is to be indicated by a meter in a 1,000-volt B+ plate lead, this high potential will exist between the current-carrying parts of the meter and the grounded panel. In this situation, the meter case should be made of insulating material, and not metal; manufacturers' recommendations should also be consulted in this respect. Most panel type meters are housed in insulated cases.

Adapting Meters. Panel type current meters of suitable current rating can be adapted to voltmeter use by means of series multipliers. When the meter is mounted upon a panel the multipliers are located outside the meter case, but must be protected against flash-over to the metal panel or low potential points which may be nearby.

In the event that a portable voltmeter such as a volt-ohm-milliammeter or a vacuum-tube-voltmeter becomes inoperative, a panel type current meter can be adapted to voltmeter use.

Most current meters contain internal shunts. These may be removed as a means of increasing the meter sensitivity (reducing full-scale current rating) in order to make the meter suitable for use as a voltmeter, with the proper multipliers based on the current rating of the movement. Removal of the internal shunt must be done with great care. Then it becomes necessary to determine the full-scale current range of the shuntless meter movement.

Fusing Meters. If possible it is sound practice to add a fuse to metered circuits. A fuse protects the meter from damage by excessively high currents. A suitable fuse rating is slightly above the value of full-scale current the meter can indicate. All but thermocouple and hot-wire meters can ordinarily stand a several hundred percent overload for a short period; consequently, a fuse rated at 100 percent higher than the full-scale meter reading can still give protection. The fuse should be placed in series with the meter.

Special Panel Meters. Panel meters with certain special features are available. For example, some are hermetically sealed to keep out harmful dust and moisture, others are constructed with special mechanical features to withstand shocks, and are classified as "ruggedized." Another special feature is the illuminated dial, in which a small panel lamp mounted inside the case lights up the scale. Extra leads are brought out to a separate source of emf for this lamp.

10-2. Volt-Ohm-Milliammeters (VOM's)

These have the basic panel-type meter movement combined with switching and adjusting devices to adapt them to read various ranges of voltage, current, and resistance, and with metallic rectifiers when intended for a-c measurements.

Typical VOM's are illustrated in Fig. 10-2. These are 1,000 ohms-per-volt and 20,000 ohms-per-volt types. The following features are generally common to this type of instrument:

Meter Movement. The movement is a standard moving-coil device with a scale especially calibrated for its various functions. Meter movements intended for this application are designed to be used in any of a variety of physical positions.

Function and Range Selector Knob. This is used to manipulate the selector switch in the instrument, which connects the various shunts and multipliers into the circuit. The selection of a-c or d-c operation is made either by the main selector knob or by a separate knob provided

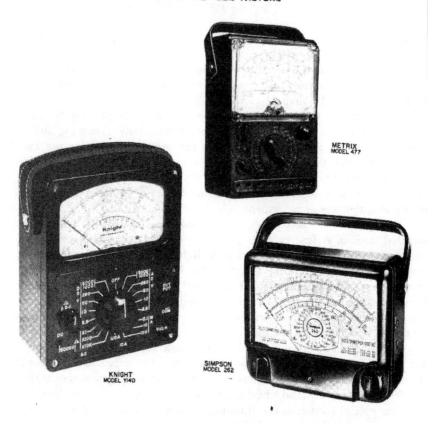

Fig. 10-2. Photographs of typical, modern volt-ohm-milliammeter. (Note: The above photos are reproduced through the courtesy of the manufacturers indicated.)

for the purpose. In the latter case, the same voltage positions are used for both ac and dc. Usually three current scales (10, 100, and 500 ma) are provided.

In some VOM's, provision is made for reading currents up to 10 amperes full scale. When measuring such high currents, be certain that the connecting leads are making good contact and are heavy enough to carry the current. The leads must prevent very low d-c resistance to avoid undue voltage drop which might disturb the circuit being tested. It is also important that the selector switch inside the instrument make good contact, or serious error may develop. The selector knob should be checked each time the instrument is used to make sure that it is securely fastened to the selector switch shaft; if it should shift, the wrong scales may be indicated and the meter overloaded or burned out.

Ohmmeter Zero Adjuster. When the selector is turned from one ohms scale to another, it is *often* necessary to reset the pointer at zero. In some cases these instruments may have two ohmmeter scales, one labeled "hi" and the other "lo" ohms. The scales run in the same direction (zero at the right) so both are of the series-type ohmmeter circuit and are zero adjusted with the test prods *shorted.* In some VOM's a shunt-type ohmmeter circuit is used for low ohms, and the pointer must be adjusted at the other end of the scale with the test prods *open.*

Tip Jacks. Some instruments use tip jacks into which the test leads are plugged. A jack labeled "COM," meaning *common,* is the one into which the negative lead is inserted. For all scales, except the high voltage or high current, the positive lead is plugged into a separate jack provided for that purpose. Special jacks for high-voltage or high-current scales generally are in a *different* location from those used for other voltage and current scales, so as to be well-insulated and isolated.

Input Resistance of Instrument. As has been explained, (Sec. 3-15 and 3-17) the total resistance of the meter on any voltage range is equal to the full-scale voltage times the meter sensitivity.

Regardless of the input resistance factor, it is always best to use the highest voltage range consistent with an indication at least one-quarter of the full scale. The order of accuracy in reading ohmmeter scales at the very high resistance end of a range is usually low because the calibrations are quite crowded. Whenever possible the range chosen should be such as to cause the meter pointer to be somewhere in the first three-quarters of the scale.

Volt-ohm-milliammeters never should be left set for ohmmeter operation; this would drain the battery. If an OFF position is provided, the selector switch should be set to OFF.

10-3. Vacuum-Tube Voltmeters

One advantage of the VTVM is its high input resistance on all ranges, of the order of 11 to 20 megohms. Some of these devices present an input resistance as high as 30 to 1,000 megohms on all a-c ranges. For a-c measurements the input impedance varies greatly among different models. Information concerning this appears in Chapter 9.

The meter movements are the same in all VTVMs, namely moving-coil permanent-magnet types, even though the ranges and kinds of scales differ. An idea of this can be had from Fig. 10-3, wherein are illustrated a variety of instruments. Among these is the center-zero scale.

As far as d-c quantities are concerned, the majority of VTVMs are designed for the measurement of voltage and resistance. Only a few laboratory type special purpose instruments provide means for direct-current measurement. On the other hand, measurement of a-c quantities is limited to voltage, although in a few instances provision is included for the measurement of capacitance. The frequency range for a-c voltage measurements extends from about 20 cps to several hundred megacycles. Only in rare instances is the upper frequency limit restricted to within the a-f band.

Operation of VTVM. In most cases the VTVM function desired is selected by means of a switch labeled SELECTOR. The range generally is chosen by means of another switch, usually of the rotary type, marked RANGE. Frequently the range selector bears markings relating to more than one function; for example, d-c resistance and d-c voltage may be shown side-by-side or above each other. Sometimes the a-c scale also is selected by the same range switch. In same cases both range and function are chosen by the use of tip-jacks into which the leads are plugged.

An example of the last mentioned arrangement is the battery operated Hickok 214 VTVM shown in Fig. 10-3. One tip-jack serves as the "hot" connection for the test lead when making d-c voltage measurements in the 3—1,200 volt range, another tip-jack serves as the input connection for a-c voltage measurements between 3 and 300 volts, and a third tip-jack is used when making a-c measurements between 300 and 1,200 volts. In contrast, the Triplett Model 650 VTVM shown in Fig. 10-4 uses a different arrangement. The a-c and r-f voltage measurements are made via a probe which plugs into a coaxial fitting. All d-c voltage and resistance measurements, on the other hand, involve a tip-jack. When a change in polarity of a d-c voltage measurement is encountered, the required reversal of the test connections is accomplished by means of a polarity selector switch marked −V and +V. In the RCA voltohmyst, shown in Fig. 10-5, polarity reversal is accomplished by the rotary selector switch.

Zero Adjustments. All VTVMs which provide for d-c resistance measurement contain two variable controls of great importance. These are the electrical-zero adjust, whereby the electrical balance of the tube circuits is attained, and the zero adjust for the ohmmeter circuit. Markings for these controls generally are the same on most instruments, although it is not inconceivable that the electrical balance zero adjust may be confused with the ohmmeter zero adjust. A distinction usually is made between these two controls by labeling the ohmmeter adjustment as OHMS ZERO or OHMS ADJUST, whereas the electrical balance control is marked ZERO ADJUST.

The ohms-zero adjustment is a simple matter, but it is important to bear in mind that it is not independent of the electrical-zero adjustment. The latter also requires adjustment when the instrument is put into use. Both the electrical-zero and the ohm-zero adjustments require that the instrument as a whole reach temperature stability in order to

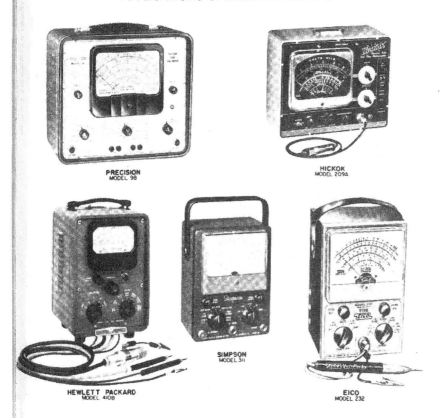

PRECISION
MODEL 98

HICKOK
MODEL 209A

HEWLETT PACKARD
MODEL 410B

SIMPSON
MODEL 311

EICO
MODEL 232

Fig. 10-3. Photographs of typical, modern vacuum-tube voltmeters. (Note: The above photos are reproduced through the courtesy of the manufacturers indicated.)

avoid the necessity for continuous readjustment. In other words, before adjusting these controls it is necessary that a reasonable amount of *warm-up* time be allowed (15 to 20 minutes warm-up time for all a-c operated devices, 5 minutes for battery operated devices).

VTVM devices are designed to be stable by making the vacuum-tube circuits highly degenerative. In spite of this, readjustment of all controls is occasionally necessary. When the instrument is set from a high range to the most sensitive range, from 1 to 5 volts, a grid current voltage drop of from .1 to perhaps .3 volt across the entire input divider can cause quite a change in the zero setting. This possibility must be watched. An increasing need for electrical-zero readjustment is a sign of improper operation developing in the instrument.

Precautions must be observed in the ohms-zero setting when changing resistance ranges. Instruments are designed for constant ohms zero

Fig. 10-4. Triplett 650 VTVM.

on the different resistance ranges, but things do happen and erroneous readings can be the result unless the zero adjustment is correct. As a safeguard against wrong resistance measurements, check the ohms-zero adjustment each time the ohmmeter range is changed. Do not take a correct zero adjustment for granted.

For d-c resistance measurements, the non-uniformity of the VTVM scales tends to limit ease of reading to the first three-quarters of the scale, although about nine-tenths of the scale is definitely usable. Whenever possible, the scale should be such as to cause the pointer to swing not higher than three-quarters of full scale.

The on-off switch of battery-operated VTVM devices should be turned off when they will be out of use for any extended period. Difficulty in zeroing the instrument is usually an indication of run-down batteries.

10-4. R-F Measurements with Probes

VTVM probes for r-f measurements are shielded and grounded, and therefore should be immune to external fields. What may be true in theory, however, is not necessarily true in practice. When measuring an r-f voltage at a point located near another source of r-f voltage, both fields may contribute to the final indication of the instrument. To minimize these effects, the probe tip length should never be made greater than provided for in the design of the instrument. Hanging an

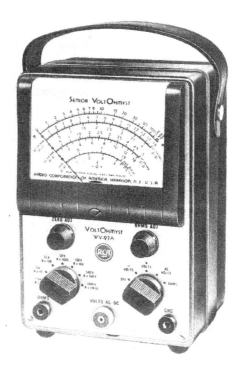

Fig. 10-5. RCA Voltohmyst.

extra length of wire onto the probe tip for the sake of convenience in measurement is not recommended. (See Fig. 10-6A.) The higher the frequency of measurement the more undesirable is this practice.

Experience also has shown the disadvantage of grasping the r-f probe head in such a way that the hand is very close to the probe tip. Body capacitance effects may introduce extraneous voltages.

A local field may produce a high and false indication when the probe tip is not in contact with the point of measurement. The free probe tip presents a high impedance across which a relatively high voltage is induced. When the probe tip is placed in contact with a test point the impedance between the probe tip and ground usually is lower, and the pointer drops to the proper indication.

R-f probes used with VTVMs provide for grounding at the point of probe contact or close by. The higher the frequency of measurement the more important it is to have a short ground connection. A sign of poor grounding is the inability to repeat a measurement with the same indication when the location of the ground connection is changed. When the ground connection issues from the probe it is important that the contact it makes with the probe be perfect. The flexing of these cables during continuous use of the probe frequently damages the wire at the point of connection; strands of multi-wire cable sometimes become

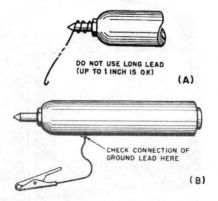

DO NOT USE LONG LEAD
(UP TO 1 INCH IS OK)

(A)

CHECK CONNECTION OF
GROUND LEAD HERE

(B)

Fig. 10-6. (A) Addition of a piece of wire to the tip of the r-f probe may introduce error. (B) A long or defective ground lead can also cause trouble.

broken, thus increasing the r-f resistance of the ground connection. The result is an unstable indication which changes with changes in probe position. (See Fig. 10-6B.)

When making r-f measurements, sine-wave voltages are assumed by the instrument manufacturer unless otherwise stated. For voltages of different waveform, even when within the frequency passband of the probe, correction factors must be applied. Such correction factors usually are furnished with the instrument if needed.

11

GENERAL ELECTRICAL TESTS AND MEASUREMENTS

11-1. Testing Batteries

Unless a battery has been dead for a long time, a no-load voltage test made by connecting a voltmeter across it is not truly indicative of battery condition. As the battery deteriorates its internal resistance increases; to detect this the test must be made by placing a load across it.

One way to check the condition is to connect a resistor of the proper value across the battery, so that the normal load current will flow through it, then apply a voltmeter. The resistor and meter can be found in one instrument called a "battery tester" (Fig. 11-1); instruments with a number of ranges for different types of batteries are available. In these, a multiple switch selects the load resistor and voltmeter range required for each type of battery. The application of a conventional voltmeter for this purpose is shown in Fig. 11-2.

A good general rule is to replace dry batteries which have dropped below 70 to 80 percent of their rated voltage under load.

11-2. Battery Charging

Figure 11-3(A) shows a simple battery-charging setup. The ammeter is connected in series with the battery to indicate rate of charge. The rate of charge depends upon the difference between the charger voltage and the battery voltage. Rates of charge actually used vary from a fraction of an ampere for small "trickle" chargers up to 100 amperes for

125

Fig. 11-1. RCA battery tester.

large "quick-charge" units. A recommended normal maximum charging current is from 5 to 10 amperes.

In some applications charging takes place while the battery is in use, as in automobiles. If, as in cars, the charging rate as well as the discharging rate varies, the net current may be charging or discharging the battery. To measure this net battery current, a *center-zero ammeter*

is used (Fig. 11-3B); if the pointer moves to the right, the charger current exceeds the lead current by the indicated amount. If the lead current is larger, the pointer moves to the left of zero.

11-3. Checking Equipment Current Drain

The circuit is Fig. 11-3C shows how to measure the current drain of single phase a-c operated devices such as electric irons, clocks, toasters, fans, radio or television receivers, etc. A rough approximation of the required range of the current meter is 1 ampere for each 100 watts of power rating of the device. In the case of irons or toasters and other devices which do not reach temperature stability immediately, the current range of the meter should exceed that calculated as stated above, because of the initial surge of current.

If the unit is d-c operated, the same circuit applies, except that the current measuring device is a d-c meter, although the a-c moving-vane or dynamometer types of meters will function satisfactorily.

A more suitable method of measuring the current drain of low-power devices which require some time to reach temperature stability, during which time the current changes, is shown in Fig. 11-3D. An a-c

Fig. 11-2. Detemining terminal voltage of
a battery.

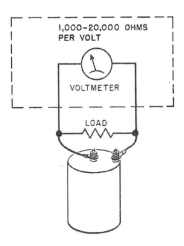

voltmeter (or a d-c voltmeter, depending on the kind of current in-volved) is connected across a series resistor and the voltage across this resistor is measured. The current drain then can be determined by the application of Ohm's Law for current or $I = E/R$ where I is in amperes, E in volts and R in ohms.

A 1-ohm resistor rated at 25 watts is rated at a maximum current of 5 amps, and a 5-ohm 25-watt resistor at about 2.5 amps. The arrange-ment shown in Fig. 11-3D is suitable only if the power rating of the device does not exceed 100 watts.

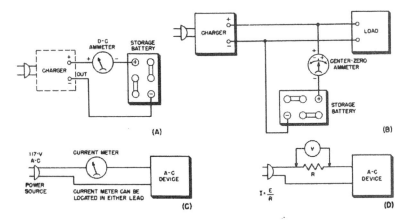

Fig. 11-3. (A) An ammeter connected to indicate rate of battery charge. (B) A center-zero meter used as a charge-discharge instrument. (C) and (D) Ammeter and voltmeter connected to measure current drain and power of device operated from the power line.

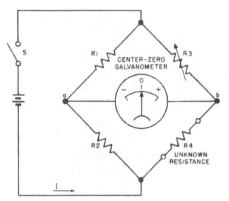

Fig. 11-4. Wheatstone-bridge circuit for measuring resistances.

FOR ZERO READING R4=R3X$\frac{R2}{R1}$

11-4.   Center-Zero Galvanometer in Wheatstone Bridge

Although most ordinary resistance measurements are made with
an ohmmeter, there are some cases in which greater accuracy than that
of the ohmmeter is necessary. The Wheatstone bridge (Fig. 11-4) provides the desired accuracy for checking and measuring such things as
shunt and multiplier resistors. *R3* is a variable resistor accurately
calibrated (frequently a decade resistance box). *R4* is the unknown
resistance.

The bridge is said to be balanced when the current *I* is divided
between the *R2* and *R4* branches in such a way that the voltage between
points *a* and *b* is zero. The center-zero galvanometer indicates this condition by reading zero when *R3* is properly adjusted. The relation given
in the diagram can then be used to determine *R4*. Resistors *R1* and *R2*
need only have a *known ratio.* They are usually selected as pairs, with
ratios in powers of 10, such as 1, 10, 100, 1,000, etc.

When a measuring system is adjusted for a balance indicated by
zero or minimum current, the balance condition on the meter is referred to as a "null." Meters used in this way are often referred to as
"null meters." A VOM or VTVM can be used as a null meter.

11-5.   Power A-F, and Supersonic Transformer Turns Ratio

Figure 11-5 shows a method for measuring the turns ratio of power
or audio transformers. An a-c voltage is applied to the primary winding.

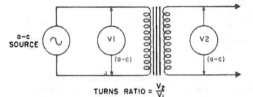

Fig. 11-5. Method of measuring
turns ratio of a low-frequency transformer. If V2 is higher than V1 the
turns ratio is step-up; if V2 is less
than V1, the turns ratio is step-
down.

TURNS RATIO = $\frac{V_2}{V_1}$

The a-c voltage across the primary winding ($V1$) and the a-c voltage across the secondary winding ($V2$) are then measured. Voltage is proportional to turns in this kind of a transformer, so the turns ratio is equal to $V2/V1$. If a heavy load is applied, the voltage ratio drops considerably below the turns ratio due to voltage drop in the windings.

11-6. Low-Frequency Impedance Measurements

A simple circuit for the measurement of any type of impedance at low frequencies is shown in Fig. 11-6 . The impedance to be measured,

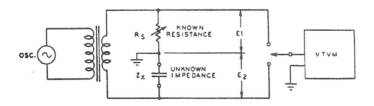

Fig. 11-6. The unknown impedance Zx is measured by comparing the voltage drop across Zx with the voltage drop across the standard resistor Rs.

Z_x, is placed in series with a non-inductive variable resistance, R_s, and connected to a source of low-frequency voltage. When the resistor R_s is adjusted so that the voltage $E1$ across R_s is equal to the voltage $E2$ across Z_x, then Z_x is equal to the value of R_s. R_s can be measured readily with a good ohmmeter or bridge, if it is not already indicated in a direct-reading resistance box. The unknown impedance also can be found from the following equation:

$$Z_x = R_s \frac{E1}{E2}$$

Because of the effect of stray capacitance, this method of measuring impedance is most suitable at power and audio frequencies. For low impedances, the method will give accurate results up to low radio frequencies.

Fig. 11-7. How the r-f impedance of a device may be determined by use of a quarter-wave line and two thermocouple r-f ammeters.

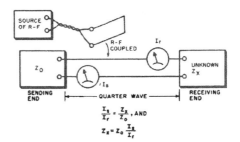

11-7. Measuring R-F Impedance with a Transmission Line

A quarter-wave transmission line can be used to measure impedance (Fig. 11-7). The line is terminatec at the *sending end* by its characteristic (surge) impedance, and the impedance to be measured

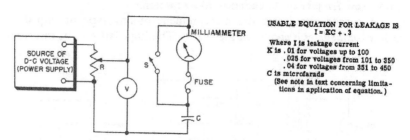

USABLE EQUATION FOR LEAKAGE IS
$$I = KC + .3$$

Where I is leakage current
K is .01 for voltages up to 100
      .025 for voltages from 101 to 350
      .04 for voltages from 351 to 450
C is microfarads
(See note in text concerning limitations in application of equation.)

Fig. 11-8. Arrangement for measuring leakage current of capacitors.

is connected at the receiving end. R-F current is then coupled into the line, and thermocouple ammeters are connected as shown. The ratio of the sending-end current (I_s) to the receiving-end current (I_R) is then the same as the ratio of the unknown impedance (Z_x) to the characteristic impedance (Z_o). The formula for Z_x is shown below the diagram.

11-8. Bridge for Measuring Impedance

Impedance of any kind can be measured by means of a bridge circuit. The principle is exactly the same for the Wheatstone resistance bridge, except that variable standard impedance and an unknown impedance replace $R3$ and $R4$ respectively (see Fig. 11-4), and the source of emf must provide a-c voltage instead of dc. The meter indicates alternating current without regard to polarity, the null being a *minimum* alternating current rather than a balance between plus and minus. For low frequencies the meter can be a rectifier type. At higher frequencies in the r-f range it must be a thermocouple milliammeter or a high-frequency probe connected to a VTVM. Note that the measurement accuracy does not depend upon meter accuracy in the usual sense, but rather upon how small a current change or minimum it can detect and on the accuracy of the standards.

11-9. Measuring Electrolytic-Capacitor Leakage Current

Leakage impairs the efficiency of electrolytic capacitors. A method for checking leakage is illustrated in Fig. 11-8. The capacitor is connected across a power supply which can be adjusted to provide the *full rated working voltage*. The milliammeter is connected in series with the capacitor, and the polarity of the applied voltage must conform with the polarity of the capacitor being tested. *The switch S is very im-*

portant and must be provided. This is because a large surge of current will flow into the uncharged capacitor when the voltage is first applied, and this surge might burn out the meter. The switch is *kept closed* until the capacitor has had about 5 minutes to charge. Then it is opened for a reading. After the switch has been opened, allow 1 minute of ageing time for each month of shelf life before checking the leakage current. The fuse is added to protect the meter in case the capacitor shorts during the test, or in case the switch should accidentally be left open when voltage is applied.

Almost any kind of milliammeter whose range is suitable for the leakage-current value is satisfactory, and sensitivity should be judged accordingly. A meter with a maximum range of 25 milliamperes is ordinarily satisfactory. The formula of Fig. 11-8 helps determine allowable leakage, which is the calculated *I* or 10 ma, *whichever is smaller.*

11-10. Testing Fixed-Capacitor Insulation Resistance (Leakage)

The insulation resistance of a fixed paper, mica, or ceramic dielectric capacitor is an important rating. Insulation resistance as low as 1,000 megohms can cause trouble if the capacitor is used as a d-c blocking element between the plate of one vacuum tube and the control grid of another.

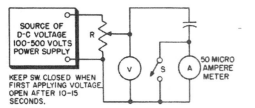

Fig. 11-9. Circuit for measuring insulation resistance of paper—mica or ceramic-dielectric capacitors.

The test circuit is shown in Fig. 11-9. Mica and ceramic dielectric capacitors should display between 5,000 and 7,500 megohms insulation resistance with 250 volts dc applied. Assuming an insulation resistance of 5,000 megohms, the current would be 0.05 microampere, hence causing almost imperceptible deflection, if any, on a 50-microampere meter. A meter deflection in excess of the barest movement of the pointer on such a meter, therefore, would be a sign of insufficient insulation resistance or excessive leakage. For other applications, as for example bypassing across relatively low values of resistance, a lower insulation resistance is allowable than for d-c blocking. The ordinary ohmmeter test is useless for such checking because the applied voltage is too low.

Paper dielectric capacitors display lower insulation resistance than mica or ceramic dielectric units. The usual rating is from 1,000 to 2,000 megohms per microfarad at rated d-c operating voltages. Investigation

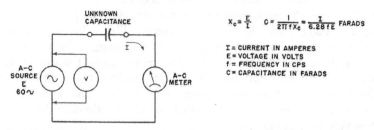

$$X_c = \frac{E}{I} \qquad C = \frac{1}{2\pi f X_c} = \frac{I}{6.28 f E} \text{ FARADS}$$

I = CURRENT IN AMPERES
E = VOLTAGE IN VOLTS
f = FREQUENCY IN CPS
C = CAPACITANCE IN FARADS

Fig. 11-10. Use of a-c source and current meter to measure capacitance. The a-c voltage source can be an isolation transformer with a 120-volt output.

of paper dielectric capacitors in perfect condition discloses that values as low as .1 μf display insulation resistances of 1,000 megohms or higher. Somewhat lower values of insulation resistance do not necessarily indicate a defective condition; the determining factor should be the leakage in the specific circuit where it is used.

11-11. Measuring Capacitance by the Current or Comparison Method

One way to measure capacitance is by its reactive effect in an a-c circuit. The greater the capacitance, the lower its reactance at any one frequency, and the greater the current which will flow in the circuit. A circuit for such measurement is shown in Fig. 11-10. The a-c meter used is a milliammeter, preferably of the multirange variety, intended for 60-cycle operation, since the voltage source generally is the 60-cycle power line. This test generally is restricted to capacitors higher than 0.5 μf, hence the lowest current scale required is about 0 to 50 ma. The current for a 0.5-μf capacitor with 120 volts ac applied is about 22 ma. For a 50-μf capacitor, with the same voltage applied, the current is about 2.2 amps.

The formula under the diagram shows how to calculate the capacitance when the current and the applied voltage in the circuit are known.

The method shown cannot be used for capacitors with an appreciable resistance component, or leakage such as electrolytics. If leakage is present, the current I is the result of this leakage as well as the capacitive reactance, and thus does not indicate capacitance alone.

A simple comparison method for measuring capacitance is shown in Fig. 11-11. The unknown is compared with a known capacitance by noting the voltage drop across a 50-ohm resistor placed in series with the capacitor. The test is made first with the unknown capacitor (C_x) in the circuit (S in upper position). Then the position of S is changed and a known capacitor from the decade box is inserted in the circuit. The decade-box selector switches are varied until substantially the same amount of a-c voltage is indicated across R for both capacitors. The unknown capacitance then has the value indicated by the settings of the selector switches on the decade box.

11-12. Measuring Inductance Capacitance by Resonance Effect

The following relation exists between inductance, capacitance, and frequency in a resonant circuit in which resistance is relatively low:

$$f = \frac{1}{2\pi\sqrt{LC}}$$

where f = frequency, cps
$\quad L$ = inductance, henries
$\quad C$ = conductance, farads

$$L_x = \frac{25.33}{f^2 C_s}$$

where f = frequency, megacycles
$\quad L_x$ = inductance, millihenries
$\quad C_s$ = capacitance, $\mu\mu f$

This relation is applied in a number of methods to measure capacitance and inductance. One of these methods is illustrated in Fig. 11-12. A standard capacitor (C_s), whose exact capacitance is known, is connected across an unknown inductance L_x whose value is to be determined. A signal from a signal generator is loosely coupled into L_x as shown, and the voltage across the parallel combination indicated by the VTVM. This meter must be suitable for the frequency at which the measurement is to be made. For any but the lower radio frequencies (below 500 kc) the high-frequency probe of the VTVM should be used.

The signal-generator frequency is then varied until the VTVM indicates maximum voltage, with a sharp drop-off on either side of the adjustment. The resonant frequency, as indicated on the signal generator, plus the known value of the standard capacitor C_s can then be substituted in the resonance formula to obtain the value of L_x.

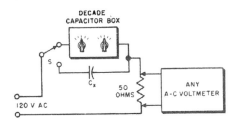

Fig. 11-11.   Comparison method for measuring capacitance.

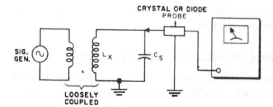

Fig. 11-12. By using the VTVM to determine the resonant frequency and selectivity of the tuned circuit, it is possible to measure inductance, capacitance, and resistance.

The same setup can be used for measuring an unknown capacitance with an inductance as the known standard. The capacitance C_s in the equation is replaced by L_s, and L_x is replaced by C_x.

11-13. Power-Supply Regulation

The voltage regulation of a power supply is of interest because it indicates how well the voltage remains constant with a varying load. The regulation of a supply can be determined by measuring the output current and voltage for various loads, as shown in Fig. 11-13. The percentage of regulation is defined in the equation just below the schematic diagram. The power rating of the variable resistance R must be adequate for the load applied. If voltage is plotted against current for different loads, a graph something like that in Fig. 11-14. results.

11-14. Power-Supply Ripple

Ripple is the a-c voltage superimposed upon the d-c voltage output. Its percentage of the total output voltage is an indication of the

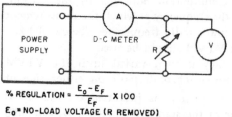

Fig. 11-13. Determining voltage regulation of a power supply.

$$\% \text{ REGULATION} = \frac{E_o - E_F}{E_F} \times 100$$

E_o = NO-LOAD VOLTAGE (R REMOVED)
E_F = FULL-LOAD VOLTAGE (AT RATED CURRENT)

efficiency of the power supply filter circuit. Figure 11-15 shows how ripple may be measured. The voltmeter (*V1*) measures the *d-c voltage* output; the voltmeter (*V2*) measures only the a-c component. The two voltages can then be used to obtain the percentage of ripple. Ripple percentage increases with increasing load current, so the test must be made at a particular desired current load, or several specified load values. The load resistor R is connected across the output, and the load current is checked with the meter *MA*. The range of *MA* must be adequate for

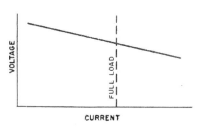

Fig. 11-14. Typical voltage regulation characteristic.

the current loads at which the tests are to be made. The capacitor C keeps direct current out of the a-c meter; its capacitance must be great enough so that its reactance is small compared to the load and meter resistances. For full-wave rectifier supplies (120-cps ripple), C should be at least 5 μf. For half-wave supplies (60-cps ripple), higher values must be used. The meter MA may be a panel meter or the direct current meter in a VOM or VTVM.

Fig. 11-15. Method of determining ripple in the output of a power supply.

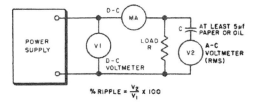

$$\% \text{RIPPLE} = \frac{V_2}{V_1} \times 100$$

12

APPLICATIONS IN RADIO AND TV RECEIVERS

Most meter applications in receivers are ones in which an external meter is employed, usually a VOM or a VTVM with suitable ranges. In general, it may be said that the primary use for meters in receivers is in *servicing* them, and the following pages are written accordingly.

12-1.   Where to Make Meter Measurements in Receivers

When trouble develops in any receiver, a variety of tests may be called for. As far as meters are concerned, the two general categories of tests are (1) d-c resistance, and (2) operating voltages. The latter may be ac as well as dc, depending on the circuit in question, although d-c measurements far outnumber a-c measurements in practice. D-c measurements embrace both current and voltage, although current measurements are not too frequent.

D-C Resistance Tests.   D-c resistance tests are made with the ohmmeter. The primary power switch on the receiver is *turned off* and the power plug is *removed* from the power line socket. Another worthwhile precaution is to allow a few minutes for the power-supply capacitors to discharge. Even then it is a good idea to discharge any and all large capacitors in the equipment being tested, by momentarily short circuiting them to ground with a screw driver.

While ohmmeter measurements can be made without removing the chassis from the cabinet, it is recommended that the chassis be removed so that all components are readily accessible.

Tube Sockets.   D-c resistance measurements can be made from the top of the tube sockets by removing the tubes and using one ohmmeter lead as the roving probe, the other lead being connected to a common reference point, usually ground or the chassis, whichever is recommended in the service notes. When individual components are checked for d-c resistance, the ohmmeter probes are connected directly

136

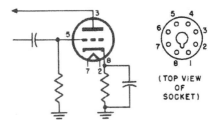

Fig. 12-1. How tube socket connection points are indicated by numbers on the schematic diagram.

(TOP VIEW OF SOCKET)

across the terminals of the component. These measurements are treated in more detail later in this chapter. Whatever is said here about using tube sockets for resistance measurements is applicable to voltage tests too.

Assuming that the chassis is the common reference point while working with the tube sockets, the following aids are suggested: First, the schematic of the receiver should be on hand for comparison purposes; this is important for learning which socket terminals are alive relative to the tube circuitry. Some receiver manufacturers use some of the free socket terminals as tie-points for attaching capacitors or resistors which may not be directly tied to active tube electrodes. The tube symbols on the schematic show which are the active tube-socket terminals by the numbers adjacent to the tube electrode symbols. This is illustrated in Fig. 12-1.

Although the top view of an octal socket shows eight holes, when there is a triode tube in an octal base only six pins (1, 2, 3, 5, 7, and 8) are used. Socket terminals 4 and 6, therefore, can be used as tie points. It is very important to realize that the schematic representation of the circuit does not show such tie points, or whether they exist. These become evident only when the underside of the chassis is examined.

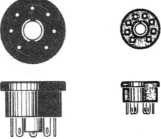

Fig. 12-2. Top and bottom views of miniature and subminiature sockets.

MINIATURE SOCKET

SUBMINIATURE SOCKET

Many octal sockets have only five active terminals (for a five-pin tube), and others make use of every terminal, because the tube inserted in the socket has eight live electrode connections. Other kinds of sockets may bear more than eight terminals, depending on the tube used, and all or only some of them may be active. Top and bottom views of other kinds, called miniature and subminiature sockets, appear in Fig. 12-2.

Contact with socket terminals for d-c resistance measurements is possible without inserting the ohmmeter probe into the socket pin holes. Adaptors with exposed connecting tongues are available for this purpose; as many tongues protrude as there are pins on the adaptor.

ADAPTOR

Fig. 12-3. An adaptor with exposed tongues.

An example is shown in Fig. 12-3. The bottom of the adaptor has pins for insertion into the tube socket, and the top of the adaptor has holes for insertion of the tube, thus permitting measurements with the tube in or out of the socket. Adaptors are available to accept almost all types of receiving tubes, even picture tubes.

All d-c resistance measurements need not be made relative to ground. When desired, these tests can be made between any two socket pins by joining the ohmmeter leads to the appropriate points, just as if they were the connections from a component.

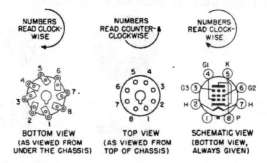

Fig. 12-4. Relation between the top and bottom views of an octal socket and schematic diagram of the tube as it would appear on a circuit diagram.

The illustrations in Fig. 12-4 correlate the bottom and top views of an octal socket. Although an octal socket is shown here, the same direction of travel when counting the pins and terminals applies to all other types of round receiving-tube sockets. Because they are being

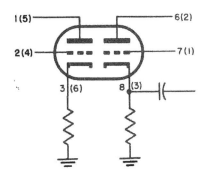

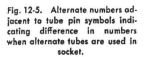

Fig. 12-5. Alternate numbers adjacent to tube pin symbols indicating difference in numbers when alternate tubes are used in socket.

viewed from opposite directions, pin numbers run in *opposite* directions in top and bottom views, although the same pin number always applies to the same pin. For applications in which an alternate tube is suggested, a receiver schematic may show two sets of pin numbers. An example is given in Fig. 12-5. The alternate tubes in this instance are a 12AU7 and a 6SN7GT.

D-C Resistance Tests on Components. Sometimes d-c resistance measurement is made with the chassis as the common reference point. Sometimes it is made directly on the component. In either case, a multi-range ohmmeter should be used and the range which affords a convenient indication chosen.

Variable Capacitors and Air and Mica Trimmers. Shorts in these components may be complete, or they may occur over only a portion of the complete range of adjustment. However, in addition to direct short circuits, high-resistance leaks may exist between terminals as the result of dirt, dust, or cracks in the insulation. Hence, a measurement capable of showing a short circuit alone is not sufficient, and measurements capable of showing leakage resistance of several megohms also should be made. Normally air and mica dielectric units will show virtually an open circuit between the active plates. When checking these units, it is recommended that they be varied through their full range of capacitance.

Fixed and Tapped Resistors. All fixed resistors are treated in a similar manner, the measurement is made between the extreme limits of the resistor in order to determine the total resistance. This is obvious if the resistor has only two terminals; but, when it has taps, an overall resistance measurement will not disclose the resistance between the taps; a connection to one of the taps may have opened up. Using one end as common, a measurement should be made at each terminal. If

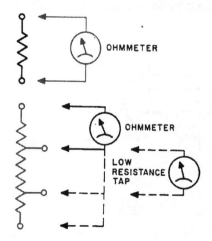

Fig. 12-6. When the resistance of a tapped resistor is being measured, each tap should be checked against a common point at one end, as indicated by this illustration. In addition, if the resistance of one section is relatively low, a separate reading directly across it is in order.

the resistance between taps is low relative to the overall resistance, individual measurement should be made between the low-resistance taps (Fig. 12-6).

When measuring the d-c resistance of receiving-type resistors used in high-frequency circuits, bear in mind the possibility of improper frequency characteristics of the resistor. Certain types of these resistors show a very great decrease in resistance to VHF and UHF currents. Yet, when measured with d-c voltage applied, they show the normal ohmic values. Therefore, make certain that the resistor is of the *proper type* relative to its frequency characteristic.

It is essential that good, firm contact between the ohmmeter probes and the measured component be made. This is particularly important at low resistance values, where contact resistance can raise the reading to far above the true component value. Tolerances in resistance values (often given on schematics) are normally from ±5% to ±20%, and can be greatly exceeded by contact resistance as above.

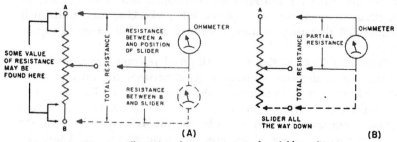

Fig. 12-7. Diagrams illustrating the measurement of variable resistors.

Variable Resistors. These are encountered in all types of electronic equipment, especially television receivers. The total resistance is measured from one end of the element to the other, as with fixed resistors (*A* to *B* in Fig. 12-7A). In rheostats, the slider must be moved to the unterminated end, as in Fig. 12-7B.

The resistance test of a potentiometer is not complete until the resistance between all positions of the moving arm and one end of the element has been observed as the arm is moved through its range. In a few special components, resistance between the arm and the element may not drop to zero even when the arm is adjusted to its minimum-resistance position; this is normal for types designed to include a small extra fixed resistance (25–1,000 ohms) as an extension of the main element.

Sometimes the moving arm of the control is purposely grounded to the case (shield and shaft) and sometimes it isn't. A continuity test between arm and case is thus in order, and the result can be checked against the schematic diagram.

Some controls do not provide for a continuous change in resistance over the full arc of travel of the moving arm. Sometimes a fixed resistance section of from 25,000 to perhaps 100,000 ohms may be a part of the potentiometer, in which case there will be no change in resistance as the moving arm is rotated over this section. Finally, the order of change in the ohmic value as the moving arm is advanced across the resistive element is not always the same, because the *tapers* are not alike. A linear taper is one in which the resistance change per degree of shaft rotation is constant as the moving arm is advanced from one limit of the resistive element to the other. Examples of different tapers are shown in Fig. 12-8.

Unfortunately, variable resistive controls do not bear labels stating the specific taper that is used. However, the function of the control usually determines the taper used, as indicated in Fig. 12-8. An abrupt change in resistance is not necessarily indicative of a fault in the unit; some controls are made that way.

In checking these controls with the ohmmeter, watch for jiggles and wiggles of the pointer as it moves across the scale. Such action indicates a bad sliding contact and a noisy control.

Resistance Test of Automatically Varying Resistors. A number of resistance components automatically vary in resistance with current through them. These cannot be checked for resistance with an ohmmeter. Instead they require the application of a fixed amount of current through the resistor and a voltage test across it. The resistance then is determined by Ohm's Law for resistance and compared with the reference information. Such units are the Globar, Thermistors, etc.

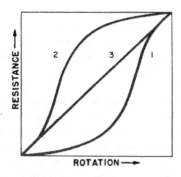

Fig. 12-8. Resistance variation
and application of various tapers
in volume controls.

1. Left-hand logarithmic taper, used in ordinary
   shunt volume controls.

2. Right-hand logarithmic taper, used in series
   controls, such as varible cathode resistors.

3. Linear taper, used in test equipment and
   elsewhere where uniform variation is needed.

D-C Resistance Measurements on Inductor-Type Components. An inductor-type component is one in which performance is based on the property of inductance. Transformers, coils, chokes, speakers, and the like are examples of inductor-type components. A d-c resistance test made on such components is of limited utility; its main purpose is to show continuity. However, under certain circumstances, it is possible to determine such things as shorted turns, or shorted layers. For such conditions, the value of d-c resistance is substantially less than the normal value, or the coil shows an open circuit.

There is no difficulty in making a d-c resistance test on an inductor type component if its nominal value of d-c resistance is 1 ohm, or a substantial part of an ohm. Measurement with an ohmmeter can indicate a defect if the fault is such as to cause a substantial change in the d-c resistance of the winding. But if the coil consists of a few turns with a relatively low total ohmic value, shorted turns may cause only a minor change in the d-c resistance of the winding, yet a major difference in the a-c performance of the device. This always has been a problem, and the solution rests in making a-c as well as d-c measurements.

Slightly lower-than-specified coil resistance could be due to shorted turns or just manufacturing tolerance (10%), in which case a resistance test is not conclusive. Substantial resistance change does, however, indicate trouble.

If a.resistor is connected across a coil, as in Fig. 12-9A, it should be disconnected when the coil is checked, unless of high enough value so that the shunting effect is. negligible.

Some circuit arrangements may call for a coil in series with a capacitor, which may not be readily evident (see Fig. 12-9B). The presence of the capacitor in the d-c resistance path will result in an open circuit indication. When the ohmmeter is correctly connected across the coil proper, it indicates the d-c resistance of that device. Measurement made across the coil and capacitor in series may show continuity corresponding to the nominal ohmic value of the coil. This is a sign of a short circuit in the capacitor.

In the case of a coil paralleled by a capacitor, measurement across the combination should show the d-c resistance of the coil. (See Fig.

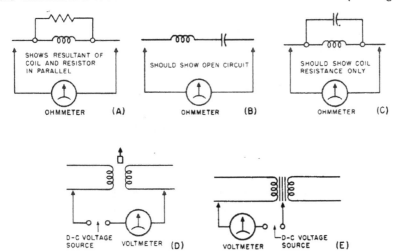

Fig. 12-9. Various combinations of coils, capacitors, and resistors which can be tested as explained in the text.

12-9C.) A lower than normal value of resistance may indicate a bad capacitor, or a bad coil. An open circuit indicates a bad coil. Another important test is to check between a coil and its core. There should be *no* leakage present.

Sometimes a d-c resistance test shows normal, yet the units seem to go bad after being placed in operation and its temperature rises. This is possible, and means that the resistance test cannot be the deciding test, but that signal or other voltage tests have to be made while the unit is in operation. This occurs frequently with horizontal output transformers and sometimes with yokes. It can also happen with an inductor unit whose temperature rises substantially during operation.

Leakage between the winding and the core, or between windings, or between a winding and a shell, is a frequent problem that must be checked. A test voltage in excess of that usually found in the ohmmeter should be used, several hundred volts d-c should be applied if possible. Many devices subject to high voltages are designed to withstand 1,500 to 2,500 volts rms between windings, and between windings and core. Electrolysis can readily cause a relatively high resistance leak between the winding and the grounded shell when the equipment is used in highly humid climates and where the salt content is high.

The methods of measuring are shown in Fig. 12-9 (D) and (E), and the insulation resistance should be up in the thousands of megohms. The voltage from a power supply approximating 300 or 400 volts is a good source of voltage; the meter can be a d-c voltmeter. The insulating resistance should be high enough so that the meter indicates zero voltage even on the lowest range.

D-C Resistance Tests on Vacuum Tubes. Frequently it is necessary to test for shorts or leaks between the electrodes of vacuum tubes, and the continuity of their filaments. Before deciding that a short circuit exists, be certain that the short circuit indication is not the result of internal connections between two electrodes. Since some tubes call for this, tube charts must be checked to verify the condition. When such tests are made, the tube should be out of its socket and the tests made at the tube base pins.

When making continuity tests on tube filaments be certain not to apply the ohmmeter voltage to low voltage-low current filaments, because they will be burned out instantly. Check 1.5 volt and 2.5 volt low current tubes by applying a low voltage with a current meter in the circuit. The applied voltage need not exceed 1 volt and the current meter can be a 250 milliampere instrument. Continuity is indicated when the meter shows current flow; its exact value is unimportant.

12-2. Point-to-point D-C Resistance Measurement

A trouble-diagnosing technique that is popular in radio receiver servicing, although it can be applied to all electronic equipments, is known as point-to-point d-c resistance measurement. In principle it is nothing more than measuring the d-c resistance between two given test points, or between a reference point and a number of different test points. The correct value of resistance for each test is known from manufacturer's or other data, either directly or from examination of the schematic diagram. If a test shows serious deviation from correct value, a defect in the circuit is indicated.

An example of the two methods of showing point-to-point d-c resistance data appears in Fig. 12-10. Sometimes the table of voltage data at the different tube socket terminals also bears the list of d-c resistance

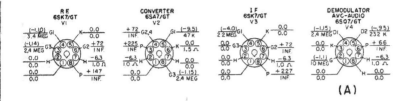

(A)

SYMBOL	TUBE	PIN 1	PIN 2	PIN 3	PIN 4	PIN 5	PIN 6	PIN 7
V-1	6BJ6	4.2 meg.	1000	22	16	500,000	1 meg.	0
V-2	6BJ6	22,000	1	30	22	500,000	500,000	0
V-3	6BJ6	20	220	38	46	500,000	500,000	0
V-4	6BJ6	4.3 meg.	120	38	30	500,000	500,000	0
V-5	12AT6	10 meg.	0	0	16	0	550,000	1 meg.
V-6	50C5	150	500,000	46	100	500,000	500,000	500,000
V-7	35W4	135	500,000	100	135	155	130	500,000

(B)

Fig. 12-10.   Bottom view of radio chassis showing d-c voltage data and resistance chart.

values between the same socket terminals and the common reference point, which is either chassis or ground.

The technique is applicable as a quick check of overall circuit continuity, or even circuit resistance when the information does not appear in chart form or on the bottom view of a chassis. In the latter case, the reference information is gathered directly from the schematic by tracing the circuitry to be checked and noting the constants of the circuit elements and their organization.

The point-to-point d-c resistance measurement technique is no different than the d-c resistance test made on an individual component, except that by selecting widely separate points as the test points all the components connected between these points are embraced at one time.

The Presence of Electrolytic Capacitors.   An important point in connection with the use of an ohmmeter for trouble diagnosis is the presence of electrolytic capacitors across resistors being checked. The positive terminal of the ohmmeter should join the positive terminal of the electrolytic capacitor. Whenever possible it is recommended that the electrolytic capacitor be removed before the circuit resistance test is made.

Still another item pertinent to successful application of the ohm-meter is that relating to the presence of selenium rectifiers. This device presents different values of resistance depending on the polarity of the voltage applied. A rapid means of eliminating the possibility of confusion because of the presence of these units is to deliberately short circuit the rectifiers by means of shorting wires.

12-3. Voltage Measurements in Electronic Equipment

Two kinds of voltage measurements may be called for as a part of the servicing routine for electronic equipment. One of these is the operating voltage, ac or dc as the case may be, and the other is the signal voltage which may be present in the system. The application of meters to·each is discussed here.

Operating Voltages. The operating voltages are of two kinds, ac and dc. D-c operating voltages exist at the cathode, control grids, screen grids, and plates of the ordinary run of vacuum tubes. If the equipment is operated from d-c primary sources, the heater or filaments of the tubes also will be subject to d-c operating voltages; if the equipment is operated from a-c primary power, the heaters or filaments are subject to a-c operating voltages.

In the event of a-c primary power operation and rectifier power supplies, a-c operating voltages will be present across the input to the power supply, across the secondary winding or windings of the power transformer, and therefore across the rectifiers, whether vacuum tube or of metallic type.

In the case of battery operated mobile equipment which uses vibrator type power supplies or dynamotors, d-c operating voltages may be present across the primary and secondary contacts of the vibrator, or only across the primary, depending on whether it is a synchronous or a nonsynchronous type. In the case of the latter, a-c usually will be present across the secondary of the transformer and across the rectifier tube. In the case of a dynamotor, d-c will be present across the motor and across the generator windings.

Operating voltages are measured between the terminals to which they are applied, whatever may be the reference point, common to all or some, of the voltages. Generally ground is this reference point. Usually it is at the same potential as the chassis or case, and is the B— point in the system. On occasion, the chassis is at a higher potential than ground, but is the B— terminal nevertheless. In almost all cases it is customary to reference operating voltages relative to ground, by which is meant B—.

Reference voltage information appears in service literature in one or more of three general forms. One is the chart form stating the operating voltages extent at the different tube electrode pins on the tube sockets, as shown in Fig. 12-11A; in another it appears alongside the tube electrode symbols on the schematic, shown in partial form in Fig. 12-11B; and, in the third form, it appears adjacent to the socket terminals for the different tube electrodes in underside chassis views, as shown in Fig. 12-11C. In all instances they bear a tolerance of plus or minus 10 to 15%.

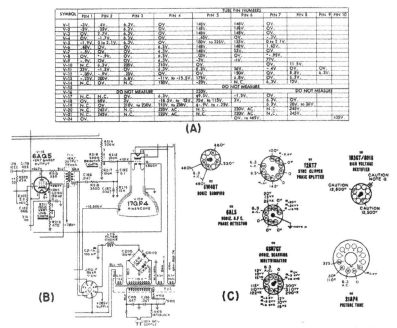

SYMBOL	TUBE PIN NUMBERS									
	PIN 1	PIN 2	PIN 3	PIN 4	PIN 5	PIN 6	PIN 7	PIN 8	PIN 9	PIN 10
V-1	-3V.	4V.	6.3V.	0V.	148V.	148V.	0V.			
V-2	-3V.	33V.	6.3V.	0V.	148V.	148V.	0V.			
V-3	0V.	7.7V.	6.3V.	0V.	148V.	148V.	0V.			
V-4	0V.	7V.	6.3V.	0V.	0V.	0V.	.3V.			
V-5	-1.9V.	0 to 2.1V.	6.3V.	0V.	180V. to 225V.	135V.	0 to 2.1V.			
V-6	48V.	0V.	0V.	6.3V.	148V.	148V.	1.65V.			
V-7	-.6V.	0V.	0V.	6.3V.	148V.	33V.	0V.			
V-8	0V.	- .9V.	0V.	6.3V.	0V.	0V.	- .95V.			
V-9	- .9V.	0V.	0V.	6.3V.	-3V.	-1V.	7.7V.			
V-10	N.C.	6.3V.	205V.	210V.	0V.		0V.	11.5V.		
V-11	33V.	-1.3V.	0V.	6.3V.	6.3V.	56V.	-4V.	0V.	0V.	
V-12	- .05V.	- .9V.	0V.	25V.		150V.	0V.	8.8V.	6.3V.	
V-13	- .25V.	200V.	6.8V.		11V. to -13.5V.	175V.	.3V.	6.3V.		
V-14	N.C.		N.C.	150V.	-20V.		N.C.	6.3V.	12V.	
V-15						DO NOT MEASURE				
V-16		DO NOT MEASURE			230V.		DO NOT MEASURE			
V-17	N.C.	N.C.	0V.	6.3V.	49.5V.	-1.3V.	0V.			
V-18	0V.	68V.	3V.	18.3V. to 13V.	79V. to 113V.	3V.	6.3V.	0V.		
V-19	N.C.	0V.	210V. to 230V.	210V. to 230V.	8-.9V. to -.25V.		6.3V.	20V. to 36V.		
V-20	N.C.	245V.	N.C.	220V. AC.	N.C.	220V. AC.	N.C.	245V.		
V-21	N.C.	245V.	N.C.	220V. AC.	N.C.	220V. AC.	N.C.	245V.		
V-24	0V.					0V. to 465V.				435V.

(A)

(B)   **(C)**

Fig. 12-11. Forms of service information: (A) voltage chart; *Emerson* (B) Partial schematic showing operating voltages next to tube electrode symbols; *RCA* (C) Voltage labels adjacent to socket terminals or underside of chassis.

Sometimes the voltage values are given in pairs in these chassis bottom views; one set of figures applies to measurements made with 20,000 ohms-per-volt voltmeters while the other set applies to measurements made with VTVMs. Naturally these are suitably labeled.

A-c operating voltages appear in the chart form and in the bottom views in the same manner as the d-c voltages. On schematic diagrams they usually are marked next to the appropriate winding of the power transformer. Sometimes a d-c operating voltage is labeled on a voltage bus which feeds a number of circuit points. If no important voltage drops intervene, the voltages at these circuit points are all equal to that at the bus. In resistance-coupled amplifiers, the voltage drops in load and coupling resistors must be taken into account. The meter must then be appropriate to the series resistance. Signal voltages in electronic equipment are essentially alternating in character. Depending on the kind of equipment, they are modulated and unmodulated sine waveforms and other complex waveform voltages.

Influence of Waveform on Voltage. Some tube electrodes are simultaneously subject to d-c and a-c voltages; sometimes the complex

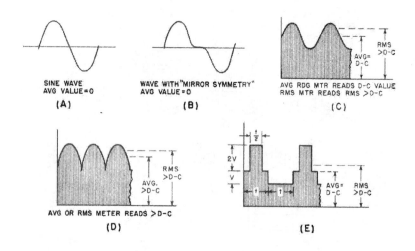

Fig. 12-12. How combinations of a-c and d-c voltages affect meter readings. A d-c voltage reading with a permanent-magnet moving-coil meter is affected by the presence of an added a-c component only if the latter has an average value different from zero, as in (D). Note, in (E), that an a-c component wave does not have to be symmetrical to have an average value of zero.

a-c voltage is symmetrical and in some cases unsymmetrical. An isolating capacitor keeps dc out of the meter for a-c measurement. A-c voltage can be kept from the meter during d-c measurements in some cases by use of a low-resistance series choke in the meter lead. Whether the a-c voltage affects the d-c voltage measurement is determined by the waveform of the a-c voltage and the kind of measuring equipment.

A wave is said to be *symmetrical* if the shape of the waveform of one polarity is a duplicate of the waveform of the other polarity. For example, a sine wave (Fig. 12-12A) is symmetrical. A special kind of symmetry is that in which the value of one polarity are repeated in *reverse order* by the other polarity, as shown in Fig. 12-12B. This is known as "mirror symmetry."

We are concerned here with measurement of voltages containing both an a-c and a d-c component. For these, the important thing about a symmetrical waveform a-c component is the fact that its *average value is zero*. Thus an average-reading d-c meter, such as the moving-coil permanent-magnet type, reads the same d-c voltage whether or not the symmetrical a-c component is added to it. If the meter movement is of the dynamometer or thermocouple type, operating on the square-law principle, it reads the rms value of the total voltage (a-c plus d-c) as shown in Fig. 12-12C. In this case, a true indication of the d-c voltage alone could only be obtained if the a-c component could be somehow removed or separated from the d-c component.

If the a-c component wave is unsymmetrical, the average value of the composite voltage (a-c plus d-c) is different than the d-c value. Then both average-reading and square-law instrument indications are affected by the presence of the a-c component with the d-c voltage.

It should be noted that an a-c component wave does not have to be symmetrical to have an average value of zero, as exemplified by the unsymmetrical zero-average a-c component of Fig. 12-12(E). Of course, such waves are rarely encountered.

As has been stated in Chapter 9, some a-c meters are intended for the measurement of sine waveform voltages, whereas others are capable of indicating the level, as peak or peak-to-peak, of complex unsymmetrical waveforms. Moreover, the measurement of a-c signal voltages also is influenced by the impedance of the source and the impedance of the voltmeter at the frequencies represented by the a-c signal, and also by the frequency of the signal and the range of frequencies over which the meter functions. This may introduce the need for supplementary equipment such as probes.

A book of this type cannot contain a discussion of the operational theory of all of the circuitry to which meter type measuring devices can be applied. It must presuppose that the user of the equipment has guidance of some sort in the behavior of the equipment under test, and is concerned with the methods involved in *applying* the different kinds of meter type measuring devices. These are discussed herein.

12-4. Suitability of the Voltmeter

The suitability of the voltmeter for making measurements of voltage is determined by a number of factors. These are:

a. The kind of voltage to be measured; whether it is ac or dc. This establishes whether the device must be a-c indicating or d-c indicating.

b. The amplitude of the voltage to be measured. This determines the required voltage range of the voltmeter.

c. The resistance of the circuit across which the voltage is to be measured. This determines the required input resistance of the voltmeter so as to minimize loading of the measured circuit and influencing the performance of the system whose voltage output is being determined. This is important in both d-c and a-c measurements. The 20,000 ohms-per-volt voltmeter has become standard as a d-c voltage measuring unit among many manufacturers of electronic equipment, although an equal number favor VTVM. The recently announced 100,000 ohms-per-volt voltmeter will, no doubt, gain popularity.

d. The waveform of the a-c voltage. This is important mainly in the self-generating systems in electronic equipment which produce complex nonsymmetrical waveform voltages which must be measured. These, and even the symmetrical a-c signal voltages, must be measured in peak-to-peak values; this requires particular design of the measuring device. Moreover, the duration of the pulse in a waveform of the non-symmetrical variety has a bearing on the type of voltmeter that is suitable for such tests.

Unless otherwise stated, all a-c measuring voltmeters are intended for sine-waveform voltages. Although the scale calibration may read in rms and peak, or even peak-to-peak values, they are not suitable for complex a-c waveforms unless specifically identified as such.

e. The frequency range of the voltages to be measured. This determines the frequency range of the voltmeter. As stated earlier, the conventional a-c voltmeter which does not use a rectifier is intended for power frequencies. Those with metallic rectifiers are intended for operation up to about 70 kc. Those with crystal or tube diode rectifiers will cover the previously mentioned range and go far above it. However, it is important to note that two factors influence the usable frequency range of even VTVMs. These are (1) the impedance of the source of voltage relative to the impedance of the meter (the higher the impedance of the voltage source relative to the impedance of the meter, the lower the frequency range of the meter), and (2) the voltage range of the meter. The constants of the device are not the same on all ranges.

Reference voltage information frequently states the kind of instrument needed. In the case of television receiver signal voltage waveforms, as well as the voltage waveforms in other kinds of equipment, it has become customary to assume that oscilloscopes will be used to indicate amplitude as well as waveform. The voltmeter cannot be expected to give the same information as the scope, but it must be realized that the peak, or peak-to-peak reading VTVM, can be used for the measurement of voltage *amplitudes* of these special waveshapes. The reference voltage figures can be gathered from the waveform patterns that appear in tv or other service literature, since the peak-to-peak amplitude values always accompany the waveform illustration.

When the reference voltage information gives the amplitude of an a-c voltage in rms values, any form of a-c voltage calibration is satisfactory, because conversion from one to other is simple. Since a sine waveform voltage input is assumed, the conversion from peak or peak-to-peak to rms values is easy. Peak-to-peak/2.83 = rms. Peak x .707 = rms. RMS x 1.41 = peak. RMS x 2.83 = peak-to-peak. But when the waveshape is complex and the peak value must be determined, then the meter must be of the type which is capable of measuring voltage waveforms of this kind; if it is not, then the conversion of rms to peak or peak-to-peak values as stated above will not be correct.

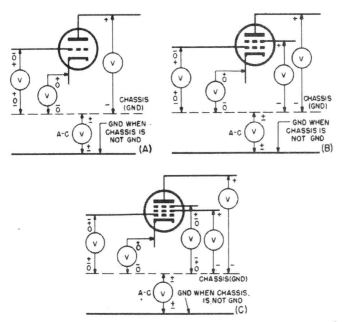

Fig. 12-13. Polarities of various vacuum-tube circuit points with respect to chassis ground.

12-5.   Kinds of Voltmeters used in Vacuum-tube Circuits

The following might be considered to be generalizations concerning the kinds of voltmeters which are suitable for d-c and a-c operating voltage measurements in different vacuum tube circuits.

a.   In control grid circuits of amplifiers, limiters, mixers, and detectors the 20,000 ohms-per-volt meter alone or as a part of a VOM is usable without any difficulty, although the VTVM will result in less circuit loading. When the voltage is *low* and the circuit resistance is *high*, the VTVM is preferred, and might be the only useful device. This is definitely so in some first a-f stages in receivers with 10 megohm grid resistors.

b.   In oscillator circuits, the VTVM is preferred for measurement in the grid circuits, and it, or the 20,000 ohms-per-volt instrument, will suffice in the screen and plate circuits.

c.   In the plate and screen circuits of amplifiers, either the 20,000 ohms-per-volt device or the VTVM can be used.

d.   All operating voltages in a-c heater circuits, low resistance cathode circuits, power line circuits, and plate circuits of power supply rectifiers, are measurable with the usual run of a-c voltmeters in VOMs and VTVMs.

e. The load circuits of diode rectifiers used as avc and agc sources will accommodate either 20,000 ohms-per-volt voltmeters or VTVMs. If the output d-c voltage is low and the load resistance is high, the VTVM will be found to perform better than the conventional voltmeter.

f. The load circuits of ratio detectors and discriminators will accommodate either 20,000 ohms-per-volt voltmeters or VTVMs.

12-6. Polarities of Operating Voltages

As far as operating voltages are concerned, Fig. 12-13 A, B, and C illustrates the general order of polarities relative to the B— point, which is assumed to be the chassis of the receiver. Triode, tetrode, and pentode types of multi-electrode tubes are considered, but the manner in which the tube is used is unimportant. Every possible condition may not be covered in these illustrations, but it is felt that the vast majority of cases are included.

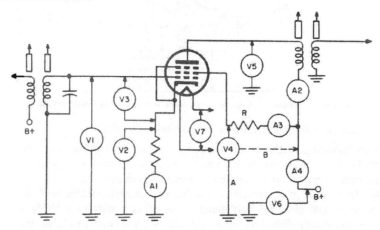

Fig. 12-14. A pentode transformer-coupled amplifier stage, showing locations for current and voltage readings. In a triode stage, readings are the same except that those connected with the screen grid and suppressor grid circuits are eliminated.

The kinds of circuit components which may be used in the grid, plate and cathode circuits of these tube structures are important relative to the operating-voltage only to the extent that they determine the resistance of the circuit across which the voltage is to be measured. It can just as readily be a resistive as an inductive load (transformer or coil) without modifying the polarity references, shown adjacent to the tube electrode symbols in Fig. 12-13, or the polarities shown at the chassis.

12-7. D-C and A-C Operating Voltage Measurement in Amplifiers

The possible points of measurement of d-c and a-c operating voltages in a transformer coupled amplifier are shown in Fig. 12-14. For the sake

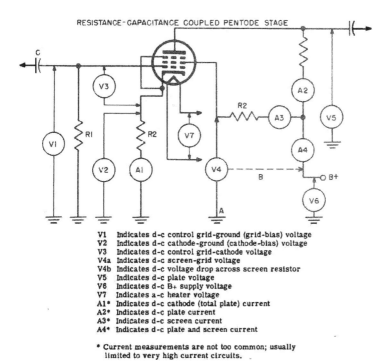

RESISTANCE-CAPACITANCE COUPLED PENTODE STAGE

V1 Indicates d-c control grid-ground (grid-bias) voltage
V2 Indicates d-c cathode-ground (cathode-bias) voltage
V3 Indicates d-c control grid-cathode voltage
V4a Indicates d-c screen-grid voltage
V4b Indicates d-c voltage drop across screen resistor
V5 Indicates d-c plate voltage
V6 Indicates d-c B+ supply voltage
V7 Indicates a-c heater voltage
A1* Indicates d-c cathode (total plate) current
A2* Indicates d-c plate current
A3* Indicates d-c screen current
A4* Indicates d-c plate and screen current

* Current measurements are not too common; usually
limited to very high current circuits.

Fig. 12-15. A pentode resistance-capacitance-coupled amplifier stage and typical meter
reading locations. For a triode in the same circuit, simply eliminate the screen grid
and suppressor circuits.

of simplification, decoupling resistors and bypass capacitors have been
omitted. The presence of these circuit components does not affect the
points of operating voltage measurement, although their manner of performance could affect the presence of the voltage, and even its magnitude.
The nature of each indication in this group is stated beneath the respective schematics in Figs. 12-14 through 12-17. The circuits are shown
with pentodes in each case. Triode circuits are handled similarly, except
that there are no screen grid or suppressor circuits.

Although it is recognized that tube current measurements are not too
common in maintenance operations, being more frequently carried out
in school or design laboratory activities, they are shown to illustrate the
points where current meters can be connected. Interestingly enough, the
use of current meters is becoming more popular in high-quality audio
amplifier systems of the push-pull type, as symbolized in Fig. 12-16.

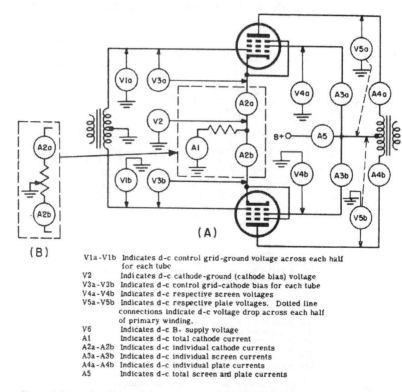

(A)

(B)

V1a-V1b   Indicates d-c control grid-ground voltage across each half
          for each tube
V2   Indicates d-c cathode-ground (cathode bias) voltage
V3a-V3b   Indicates d-c control grid-cathode bias for each tube
V4a-V4b   Indicates d-c respective screen voltages
V5a-V5b   Indicates d-c respective plate voltages. Dotted line
          connections indicate d-c voltage drop across each half
          of primary winding.
V6   Indicates d-c B₊ supply voltage
A1   Indicates d-c total cathode current
A2a-A2b   Indicates d-c individual cathode currents
A3a-A3b   Indicates d-c individual screen currents
A4a-A4b   Indicates d-c individual plate currents
A5   Indicates d-c total screen and plate currents

Fig. 12-16. Push-pull transformer coupled stage, showing possible d-c operating
voltage and current readings. For triodes, eliminate screen grid and suppressor
grid circuits.

Individual current meters are located in the cathode circuits, or at least jacks are provided for them so that the cathode currents in the individual tubes which form the push-pull stages can be adjusted for proper balance. The arrangement used for cathode bias, to omit this conveniently, is shown in the respective dotted line enclosures. By varying the slider on the potentiometer type of cathode bias resistor, variations in the tube currents can be compensated for.

The phase inverter system shown in Fig. 12-17 is but one example of a great many. The manner of obtaining the phase inverted signal from the single-ended input tube differs in many systems; but even so this circuit illustrates the numerous points where d-c and a-c operating voltages are applied so as to form a system which will result in a balanced input to the output stage and a similar output. It is interesting to note the organization of the phase-inverter tube operating voltages, especially the use of the grid resistor *R4* system of one output tube as the grid resistor for the phase-inverter tube.

Grid Voltage Measurements. Amplifier circuits which use blocking capacitors in the control grid circuits raise an interesting point relative to d-c bias voltage measurements in these circuits. Examples of this are the voltmeter readings *V1* and *V3* in Fig. 12-15, and *V3* and *V6* in Fig. 12-17. Excessive leakage in the blocking capacitors in control grid circuits is not uncommon. When this occurs the preceding tube plate voltage causes current to flow though the grid resistor of the tube in the succeeding stage. Thus, in Fig. 12-15, such action would result in a current flow through *R1* and develop a voltage drop across it. This would be indicated on *V1*. Under normal no-signal conditions this meter would indicate zero voltage.

In the event of leakage in the blocking capacitor *C*, the normal bias developed by the cathode resistor *R2* is partially (or completely) overcome by the current through *R1,* and the *V3* indication is less than the *V2* indication. Normally these are alike, except possibly for what slight difference might be occasioned by the presence of *R1* (if it is very high) in the voltage measuring circuit.

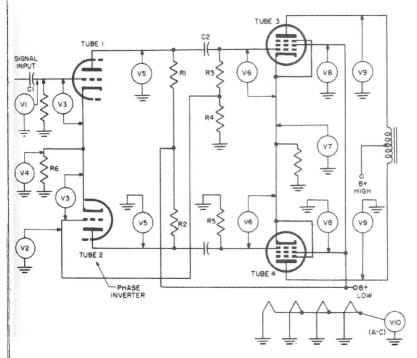

Fig. 12-17.   Possible d-c operating voltage measurements in a typical phase inverter push-pull amplifier system.

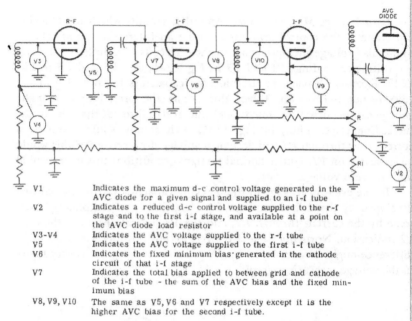

V1   Indicates the maximum d-c control voltage generated in the
           AVC diode for a given signal and supplied to an i-f tube

V2   Indicates a reduced d-c control voltage supplied to the r-f
           stage and to the first i-f stage, and available at a point on
           the AVC diode load resistor

V3-V4   Indicates the AVC voltage supplied to the r-f tube

V5   Indicates the AVC voltage supplied to the first i-f tube

V6   Indicates the fixed minimum bias generated in the cathode
           circuit of that i-f stage

V7   Indicates the total bias applied to between grid and cathode
           of the i-f tube - the sum of the AVC bias and the fixed min-
           imum bias

V8, V9, V10 The same as V5, V6 and V7 respectively except it is the
           higher AVC bias for the second i-f tube.

Fig. 12-18.   Examples of d-c control voltage measurements in radio receiver.

A leak between plate and grid winding in a transformer-coupled system, such as in Fig. 12-14, can also be harmful. It can modify the control grid-cathode voltage shown on *V3*, or it can overcome it completely. In the last event the control grid would be positive and the indication on *V1* would be a positive voltage representing the voltage drop developed across the secondary winding of the transformer.

The significance of these few circuit examples does not lie in the specific procedure, but rather in the philosophy that the operating requirements of the tube dictate the application of operating voltages. Although we have remarked that operating voltages are relative to a B— reference point, we have given several examples to show a departure from this. These are the control grid-cathode voltages, which are important because they show that the control grid circuit voltage can be zero rather than negative, depending on the reference point.

12-8.   D-C Voltage Measurements in AVC Circuits

The measurement of d-c voltages in control circuits warrants special comment. Inasmuch as the normal d-c and a-c operating voltages are measured without signal input, it is important to note that control voltages are absent unless a signal is applied to the system.

The usual function of a control circuit is to furnish an automatically generated and varying control grid bias which varies the behavior of the amplifier tubes. Sometimes this control voltage is the only bias applied to a tube, as for example in the r-f stage shown in Fig. 12-18. At other times it is an added bias applied to a fixed minimum generated in the cathode circuit, as shown in the i-f stages in the same illustration.

The types of control circuits used in radio receivers vary in specific organization, but do bear great resemblance to one another.

The amount of bias voltage generated in the avc diode (or triode as the case may be) is a function of the applied signal. This makes possible using the avc voltage as an indicator of the condition of resonance when aligning a receiver. In some circuits, called "delayed avc" circuits, avc bias is zero unless the received signal voltage exceeds a predetermined minimum value.

12-9. D-C Voltage Measurements in Oscillators

Measurements of d-c operating voltages in oscillators used in radio, tv, and other equipment depend on the function of the oscillator. If it is a heterodyning oscillator, the d-c voltage measurements generally are two in number: the rectified grid voltage, generated while the tube is oscillating, and the plate voltage. These are shown in Fig. 12-19A and B for two common kinds of oscillators. These need not be triodes, they can be other tube types.

As a rule, the plate voltage measurement ($V2$) in both illustrations is less important than $V1$. If the latter shows the control grid to be negative relative to ground, the system is oscillating; and if the magnitude of the voltage measured or $V1$ conforms with the reference information, the $V2$ measurement is unnecessary. The magnitude of the rectified grid voltage is proportional to the amplitude of oscillations. Grid voltage must always be measured with a high-impedance meter, preferably a VTVM, to minimize loading on the high-impedance grid circuit.

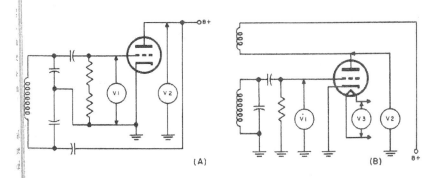

Fig. 12-19. Examples of operating voltages: (A) d-c in Colpitts oscillator; (B) d-c and a-c in tickler feedback oscillator.

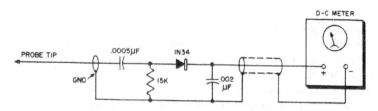

Fig. 12-20. Typical circuit for r-f probe, which is sensitive to the field of an oscillator without contacting the oscillator circuit.

Although only two examples of heterodyning oscillators are shown here, all other types are treated in a similar manner. The control grid will be negative relative to ground when the system is oscillating. The amount of voltage present across the grid leak will vary with the conditions in the circuit, and with the frequency setting if it is a variable frequency oscillator. Practical values can be as low as a few volts to as high as 20 or 30 volts.

Measurement of the plate voltage or grid voltage may cause a change in output frequency. This is not harmful because it exists only while the meter is connected across the circuit. However, if the loading effect of the meter is enough to stop oscillation the check is worthless.

The field surrounding an oscillating system affords a simple means for determining if an oscillator is functioning. Detection of this field by probe pickup without contact permits the use of a low input resistance meter without undue loading. The circuit is shown in Fig. 12-20.

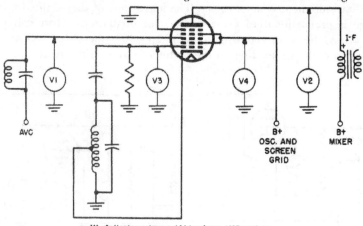

V1 Indicates mixer grid bias from AVC system
V2 Indicates mixer plate voltage
V3 Indicates oscillator grid voltage
V4 Indicates oscillator plate (and screen grid) voltage

Fig. 12-21. Typical converter circuit with voltage-checking points.

An ordinary crystal probe is connected to a meter which is set to a low d-c voltage range. The probe tip is held *near* the elements of the oscillating circuit, near the tube grid terminal at the bottom of the socket, or at its cathode, or near the tank coil, but *not* in contact with these points. If the circuit is oscillating, the crystal in the probe will rectify the field energy and a voltage will be indicated on the meter.

12-10. D-C Voltage Measurement in Converters

Measurement of d-c and a-c operating voltages in receiver converters presents no peculiar problems. They are made in the same way as shown in Sections 12-6 through 12-11. Converters are single-envelope combinations of mixers and oscillators, hence the possible points of operating voltage measurement in the multi-electrode tubes used for this function represent the composite of the section used as the mixer and the section used as the oscillator. Since only the cathode is common to both sections, a single cathode-ground voltage measurement will suffice.

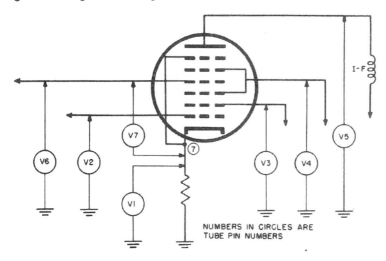

NUMBERS IN CIRCLES ARE
TUBE PIN NUMBERS

V1 Indicates cathode bias active for the oscillator and
   mixer sections
V2 Indicates oscillator grid (pin 4) voltage
V3 Indicates oscillator plate (pin 3) voltage
V4 Indicates mixer screen grid voltage
V5 Indicates mixer plate voltage
V6 Indicates mixer control grid-ground voltage
V7 Indicates mixer control grid-cathode bias voltage

Fig. 12-22.   Another type of converter circuit.

A typical converter circuit is shown in Fig. 12-21. It is a 6SA7 pentagrid tube. One section is grids 1 and 2 acting in conjunction with the cathode form the oscillator electrodes. Grids 3, 4 and 5 and the plate, acting with the cathode, form the mixer section. Grid 4 is the screen grid and grid 5 is the suppressor grid. The connections between grids 2 and 4 are made internally.

As far as d-c operating voltages are concerned, four electrodes receive d-c voltages. The oscillator grid 1, the oscillator plate 2 (and the screen grid 4), the mixer grid 3, and the plate. These are labeled in Fig. 12-21.

Another version of a converter, the 7A8, uses six grids, a cathode, suppressor grid, and plate. Its name is octode converter, and it is shown in Fig. 12-22.

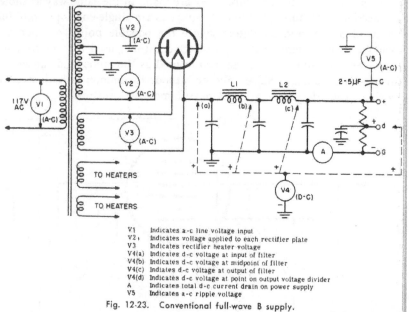

V1	Indicates a-c line voltage input
V2,	Indicates voltage applied to each rectifier plate
V3	Indicates rectifier heater voltage
V4(a)	Indicates d-c voltage at input of filter
V4(b)	Indicates d-c voltage at midpoint of filter
V4(c)	Indiates d-c voltage at output of filter
V4(d)	Indicates d-c voltage at point on output voltage divider
A	Indicates total d-c current drain on power supply
V5	Indicates a-c ripple voltage

Fig. 12-23.   Conventional full-wave B supply.

The precautions stated in connection with d-c operating voltage measurements on oscillation applies to the oscillator section of these dual functioning tube systems. It also should be realized that an inoperative oscillator section will not prevent the presence of d-c operating voltages on the mixer section.

12-11. Measuring Voltages in Power Supplies

A variety of power supplies are used in electronic equipments. They are classified as low voltage and high voltage. The low voltage supplies furnish outputs of from a few volts to several hundred, while high voltage power supplies furnish thousands of volts output.

Two power supplies are shown in Figs. 12-23 and 12-24. Both are full-wave rectifiers that typify low voltage systems. Dealing with Fig 12-23 first, the primary voltage input to the power supply is measured across the primary of the input power transformer. This is an a-c voltage and requires an a-c voltmeter (*V1*). The voltage fed to the rectifier tubes for conversion to pulsating dc and subsequent filtering also is of an *alternating* character. It can be measured across each half of the high-voltage winding. The center tap can be a common point for these measurements (though not for the primary-voltage measurement). These meter connections are shown as *V2*. The heater voltage fed to the rectifier also is ac and secured from another winding on the transformer. An a-c voltmeter of suitable voltage rating is connected as is *V3*.

All of the a-c measurements on a power supply are made across relatively low resistance circuits, hence any conventional a-c meter of suitable accuracy and frequency rating can be used. The rectifier type is preferred. Since the output of the rectifier is a pulsating d-c voltage, it will be indicated on a d-c meter.

Since the power supply has a common ground for the entire filter system, ground can be the common reference point. All other points will be positive relative to ground. Therefore the negative terminal of

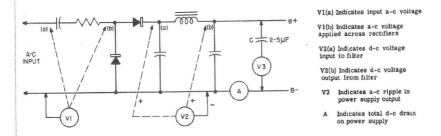

Fig. 12-24.   Selenium rectifier power supply (full-wave voltage doubler).

the voltmeter *V4* is joined to ground, and the voltages at different points in the filter system, and at the output, are determined by the roving positive lead. If the power supply contains a half-wave rectifier, voltage measurement proceeds as outlined above, except that fewer a-c measurements are made.

A simple full-wave selenium rectifier, "transformerless" voltage-doubler power supply is shown in Fig. 12-24. The a-c voltage measuring points are shown by *V1*, and the d-c voltage measuring points are shown by *V2*. The ripple across the output is measured by connecting a low-voltage meter of the rectifier type in series with a 2 to 5 μf capacitor across the output of the power supply, as shown by *V3*. The working voltage rating of the capacitor *C* should be twice the power supply output voltage.

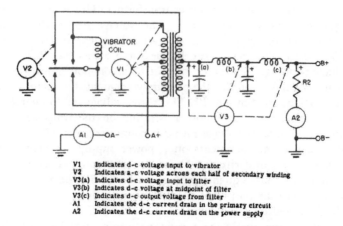

V1	Indicates d-c voltage input to vibrator
V2	Indicates a-c voltage across each half of secondary winding
V3(a)	Indicates d-c voltage input to filter
V3(b)	Indicates d-c voltage at midpoint of filter
V3(c)	Indicates d-c output voltage from filter
A1	Indicates the d-c current drain in the primary circuit
A2	Indicates the d-c current drain on the power supply

Fig. 12-25. Full-wave synchronous vibrator power supply.

Numerous devices, especially mobile electronic equipment operated from a d-c source, use vibrator type power supplies. One example is given in Fig. 12-25; this is a full wave synchronous unit. The vibrator does the rectifying by being active in both the primary and secondary circuits of the transformer. Adjustment of the rectifier is usually done while using a scope, nevertheless voltage tests are useful as a rapid means of determining performance.

A nonsynchronous vibrator type of power supply differs from the synchronous type in that the vibrator is used only to interrupt the primary current. Rectification is done by a full-wave rectifier tube.

Dynamotor Power Supplies. Mobile electronic equipments occasionally use dynamotor power supplies. These are small motor-generator units operated from 6-32 volt storage batteries, and furnishing d-c output at various voltage up to about 500 volts. They differ from the usual motor-generator unit in that a common field winding is employed for both. Schematically the dynamotor power supply appears in Fig. 12-26.

The application of voltage and current measuring devices is simple as indicated. Since all the circuits involved have low resistance, sensitivity of the voltage measuring devices is not a factor. The measurements should be made under rated current loads or otherwise the indications do not reflect the true performance of the device.

12-12. Measuring Operating Voltages in TV Receivers

The measurement of operating voltages (both dc and ac) in television receivers is not unlike that in radio receivers or other electronic equipments which were described in Sections 12-5 through 12-10.

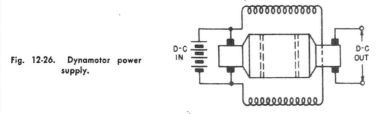

Fig. 12-26. Dynamotor power supply.

However, one particular portion of a television receiver requires special techniques and equipments for the measurement of the d-c voltages in the presence of a high a-c voltage which exists in it. This portion is the horizontal-output tube plate circuit, the horizontal-output system, and the high-voltage supply. These are symbolized in Fig. 12-27.

All horizontal-output systems may not be exactly like the one shown here, but the points indicated will be found in virtually all systems, hence the comments apply.

The plate (7) of the horizontal-output tube is subjected to a d-c operating voltage of from 250 to 500 volts. As a rule this voltage exceeds that available from the low voltage B+ supply (4) by an amount corresponding to the "boost" voltage (5) developed in the damper systems across the "boost" capacitor.

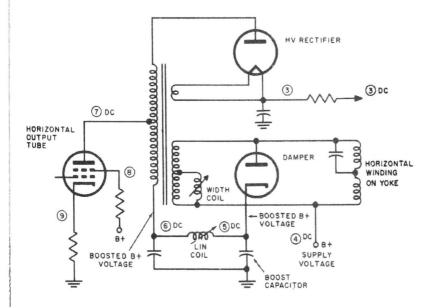

Fig. 12-27. Typical horizontal output system and high-voltage supply.

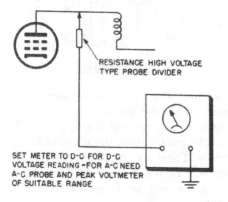

ANY TUBE AT WHICH HIGH VOLTAGE A-C
SIGNAL AND PLUS OPERATING D-C VOLTAGE
EXISTS AT THE PLATE OR ANY OTHER TERMINAL

RESISTANCE HIGH VOLTAGE
TYPE PROBE DIVIDER

Fig. 12-28. Measuring d-c operating voltages where both a-c and d-c voltages are present.

SET METER TO D-C FOR D-C
VOLTAGE READING –FOR A-C NEED
A-C PROBE AND PEAK VOLTMETER
OF SUITABLE RANGE

At the same time, the horizontal-output tube plate is also subjected to a very high a-c pulse of nonsymmetrical waveform. Hence two voltages exist simultaneously at this point.

The measurement of the d-c operating voltage at the horizontal tube plate or at any point in the horizontal-output system where both d-c and a-c voltages exist can be accomplished as shown in Fig. 12-28. The meter can be either a VOM·or a VTVM. A high resistance d-c voltage divider probe with known voltage ratio is required between the horizontal tube plate and the meter. Only a fraction of the voltage at the probe tip appears across the meter. As a rule these probes afford about 100:1 ratio in voltage division. The meter should be adjusted to the 10 volt d-c scale. Using a probe of this kind will make the high a-c pulse voltage at the horizontal-output tube plate ineffective as far as the d-c meter is concerned. The operating d-c voltage is equal to the amount shown on the meter times the voltage divider ratio of the probe. The test point for this measurement is (7) in Fig. 12-27.

Other test points for d-c voltage measurements in horizontal-output systems are the screen grid (8) and the cathode (9) of the output tube; the low end of the horizontal-output transformer primary (6) in Fig. 12-27, the cathode of the damper tube (5) and the B+ supply feed point (4). The divider probe is not required at these points, although no harm comes from using it; the VOM or the VTVM can be used directly. The voltages are approximately from 250 to 500 dc.

Sometimes the horizontal-output circuit is such that the plate of the damper tube is the low side. In this event the point (5) in Fig. 12-27 would be at the plate of the damper rather than at the cathode. Measurements at points corresponding to (6) or (5) and (4) indicate the unboosted and boosted B voltage for the horizontal-output tube. Points (6) and (7) would be subject to the same d-c voltage.

The high-voltage d-c probe is made of special insulating material, and is of special construction to afford protection against electric shock. One example of such a high-voltage probe is shown in Fig. 12-29. It contains a network of high-value resistors, and only a small fraction of the total voltage applied across the divider appears across the voltmeter.

Another application for the high resistance voltage-divider probe is the measurement of the d-c output of the high-voltage power supply, which is the same as the second anode voltage. This too is shown in Fig. 12-27. The test points are labeled (3). Unlike the horizontal-output tube plate, where d-c voltages of from 250 to perhaps 500 volts are found, the d-c voltage at the output of the high-voltage power supply approximates from 8,000 to perhaps 18,000 volts, depending on the specific receiver. Hence the voltmeter range for a 100:1 divider probe should be the 250 or 300 volt scale in order to protect the instrument.

Fig. 12-29.   Example of a high-voltage probe.

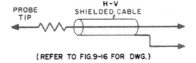

(REFER TO FIG.9-16 FOR DWG.)

High voltage measurements must be made very carefully, using a high voltage divider probe. It is a good idea when measuring voltages at high voltage points in a tv system to turn off the receiver first, connect the probe to the test point, and then turn the receiver on again.

Operating Voltages in Blocking Oscillators and Multivibrators. The measurement of d-c operating voltages at tube electrodes in blocking oscillators and multivibrators in television receivers is conventional, despite the fact that a-c signal voltages also are present. The a-c voltages at these points are not too high, and the fact that they approach a symmetrical waveform makes it possible to measure the d-c voltage without concern for the a-c which is present.

Operating d-c voltages which are of interest in blocking oscillators are the plate voltage, the control grid voltage, the cathode voltage, the a-c heater voltage, and such other voltage as may be listed in the reference information. The same is true of horizontal oscillators and multivibrators. It is essential to bear in mind that the grid, plate, and cathode electrode operating voltages encountered in these systems are subject to variations according to the setting of the associated controls.

Spurious Signal Pickup in TV Receivers. Spurious signal pickup with a rectifier probe will be plentiful in a television receiver. The strong field from the horizontal-output system will result in signal indications almost regardless of where the probe tip may be located. Hence, when attempting to make measurement with a rectifying probe or a peak-to-peak probe in a television receiver, it is advisable to first remove the horizontal oscillator tube, provided that this is not the system which is being tested.

When checking pulse amplitudes in the vertical integrator system it is advisable to remove the vertical oscillator, because signals from this tube find their way into the vertical integrator.

Adjusting 4.5-MC Trap. The crystal or diode probe used with a VTVM, or possibly with a VOM having a sufficiently low d-c scale (0–3 volts), can be used for the adjustment of traps if the probe tip is connected to points where a suitable signal exists. For instance, in inter-carrier television receivers, the 4.5 mc sound signal is not routed in the video system. Therefore the rectifier type probe voltmeter combination can be used in the video system when adjusting the 4.5-mc trap. The point of connection can be at the video signal input to the picture tube, provided that the peak voltage to ground at this point does not exceed the working voltage rating of the isolating capacitor in the probe. The meter indication should be *minimum* when the sound trap is properly adjusted. The 4.5-mc test signal is secured from a marker generator and fed into the video system ahead of the 4.5 mc trap.

Care must be exercised when using r-f probes at plate circuits of i-f amplifying tubes. The blocking capacitor in the probe has a peak voltage rating which should not be exceeded by the operating plate voltage applied to the tube, if damage to the capacitor is to be avoided. Rectifying probes designed for + voltage output will give reversed readings if the isolating capacitor has excessive leakage and the probe is joined to a source of dc.

12-13. Alignment and Other Tuning Adjustment

The VOM and the VTVM are frequently used as indicators for peak type adjustments on tuned circuits, r-f, mixer, oscillator, and i-f coils and transformers, in radio, television, and other types of receivers. They are located at points where, as the result of rectification in the electronic system, a d-c voltage is available, the magnitude of which voltage reflects the correctness of the tuned circuit adjustments.

The usual location of such indicating meters is across the second-detector load circuit in radio receivers, and across the video-detector load circuits in television receivers. In the latter instance it can be at the point of origin of the agc voltage. Another location is across the limiter grid resistor.

When the indicating meter is a 20,000 ohms-per-volt instrument and is connected across a grid resistor in a tuned circuit, it is a good idea to use a 100,000-ohm isolating resistor in series with the ungrounded

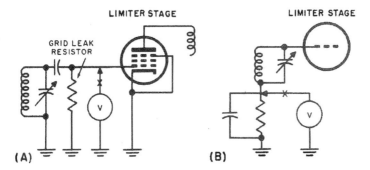

V indicates grid voltage developed by limiter
grid current. Adjust for maximum meter indication. Point X is location of isolating resistor for VOM.

Fig. 12-30. Peak alignment of a limiter stage.

lead. Examples of this are shown in Fig. 12-30. Such a resistor is not
needed when a VTVM is used because the usual d-c probe used with
VTVMs contains a 1-meg isolating resistor.

When using a VOM or a VTVM as an indicating device during
tuned circuit adjustments, it is a good idea to disable the avc and use
a fixed bias of several volts instead. The receiver manufacturer's recommendations in this connection take precedence over anything said here.

12-14. Static and Dynamic Measurements on A-F Amplifiers

A-f amplifiers are found in all kinds of electronic devices, radio
receivers, television receivers, public address systems, high fidelity repro-
ducing systems, and others.

Meters are applied to these systems for two main purposes: (1)
static measurements to check operating voltages and currents, and (2)
to measure signal voltages. The oscilloscope is the preferred device for
signal tracing because it shows waveform, but meters are also extensively
used for signal measurement. As to meter devices used for this work,
VTVMs are preferred for all high resistance control grid circuits; 20,000
ohm-per-volt instruments are usable but will not be as correct in their
indications as VTVMs. In plate, screen, and cathode circuits either
one will do.

Most a-f amplifier stages are resistance-coupled except for the output stages. The schematic in Fig. 12-15 shows a typical resistance-coupled
stage while Fig. 12-16 shows a class-A push-pull stage. The principal
points at which voltage and current readings can be made are shown.

Inasmuch as a push-pull system is supposed to be balanced to d-c
and a-c voltages, the paired voltages V_3 and V_4 (Fig. 12-16) should

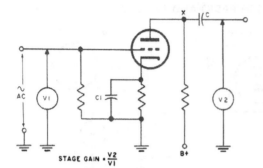

Fig. 12-31. Arrangement for measuring stage gain of an a-f amplifier using a-c meter of suitable type.

STAGE GAIN $= \frac{V2}{V1}$

be alike. Different readings indicate a difference in d-c resistance in the transformer windings, or unlike tube constants. The latter is particularly true in the case of the cathode voltage readings, which depend in great measure on the voltage drops across the two halves of the output transformer primary.

The push-pull stage shown in Fig. 12-16 can be a Class B system, or a Class AB system, in which event grid current flows during a portion of the input-signal voltage cycle. With signal input, the voltmeter $V1$ indication will not be zero voltage, nor will the readings on $V3$ be the same with or without signal voltage applied. Finally, $V3$ will not necessarily read the same as $V2$. The $V1$ indications will depend on the voltage drop developed across each half of the input tranformer secondary by the grid current. In like manner, the difference between the $V2$ and $V3$ indications will depend on the amount of grid current.

12-15. Stage Gain

The gain in an a-f amplifier stage can be determined by measuring the signal voltage at a standard frequency, such as 400 cps, at the stage input and output, inclusive of the coupling system. The circuitry for an r-c coupled stage is shown in Fig. 12-31. The signal is measured with an a-c voltmeter ($V1$) of suitable input resistance. In view of the high value of grid leak usually used in these circuits, it is best if the measurement is made with a VTVM. The output signal is measured by $V2$. If the measuring instrument contains a blocking capacitor, the high side of the voltmeter can be connected at X, the top of the load resistor. If it does not incorporate the blocking capacitor, then the measurement should be made after the blocking capacitor C. The calculation of the stage gain is shown by the equation under the schematic in Fig. 12-31. This is simply the ratio between the output and input signal voltages, and is the same for other type tubes.

A method of measuring gain in a transformer-coupled stage is shown in Fig. 12-32(A). The points of measurement are the same as for

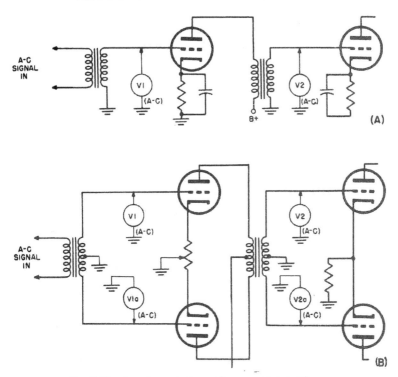

Fig. 12-32. Measuring gain in transformer-coupled amplifier.

the resistance-coupled stage, and the equation for stage gain is the same. When this test is applied to a transformer-coupled push-pull system, the gain is measured as the ratio of plate-to-plate signal voltage to grid-to-grid signal voltage. See Fig. 12-32(B).

12-16. Measuring Power Output

The arrangement shown in Fig. 12-33 is suitable for measuring the power output of any type of a-f amplifier. Output signal voltages can be measured across the loudspeaker voice coil, but uncontrollable variations in the acoustic loading of the speaker make it difficult to ascertain what the actual, resulting electrical voice coil load resistance is. Therefore, a non-inductive *load resistor* R is usually used to replace the loudspeaker or other output device. R has a value equal to the nominal impedance of the load.

Any rectifier-type meter or VTVM which measures a-c voltage at the desired frequency is satisfactory for this test. Since R is usually a replacement for the voice-coil impedance and is normally from 3.2 to 15 ohms, high meter sensitivity is not required.

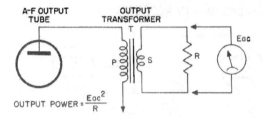

Fig. 12-33. Measuring the power output of an a-f amplifier.

12-17. Frequency Response

The frequency response of an amplifier is usually expressed on a graph on which output is plotted against frequency, with the input signal voltage to the amplifier constant. The input signal level is kept at some value which provides rated power output, or some specified fraction thereof, at a standard frequency of 400 or 1,000 cps.

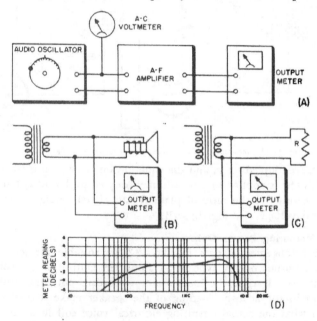

Fig. 12-34. Arrangements for making frequency-response measurements of an a-f amplifier.

Figure 12-34 shows the arrangement. Various values of frequency covering the expected range are successively used, and the output for each noted. An example of a frequency response curve developed in this manner is shown in Fig. 12-34D.

12-18.   Diode Checking by Resistance Measurement

Since the usefulness of a crystal diode, like that of any other rectifier, depends upon its ability to pass current in one direction much better
than in the other, its condition can be tested by comparing the forward
and reverse resistances, as measured on an ohmmeter scale of the VOM.
Although it is possible to perform this test without determining the
ohmmeter polarity, a much better understanding of the performance
of the test meter can be obtained if the polarity of the ohmmeter is
first definitely established, since it might differ between instruments of
different manufacture. This is best done with the aid of another d-c
voltmeter, so that the nature of the battery supply of the ohmmeter
for each of its ranges can be established at the same time. To take a
particular example, it might turn out that the black test lead of the
ohmmeter turns out to be the one that is internally connected to the

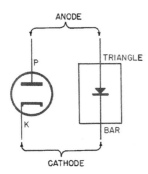

Fig. 12-35. Crystal-diode symbol corresponding to tube-diode symbol.

plus of the ohmmeter battery. In this case, the black lead must be
applied to the anode of the diode to obtain its forward resistance, provided, of course, that the battery voltage that corresponds to the particular ohm range selected is not excessive. Generally speaking, it is
safe to use the R × 100 range, since this usually will apply only 1.5 or
3 v to the diode under test. It is best to avoid the extreme high-resistance
ranges since they often apply up to 30 v, which may be beyond the
reverse-voltage rating of the diode.

Much confusion has existed in the past regarding the matter of
identifying the so-called positive terminal of a rectifier in general; (that
is, the terminal to which the positive side of the battery circuit must be
applied to obtain forward conduction). This conclusion can be avoided
by using the terms "anode" and "cathode" for the two terminals of the
diode, so that they have the same meaning as the corresponding plate

and cathode of a tube diode. This correspondence in symbols is illustrated in Fig. 12-35.

In the symbol for the crystal (or semiconductor) diode, the triangu-
lar portion corresponds to the tube anode, and the bar portion corresponds to the cathode. This is all that is needed to keep any diode straight,
no matter where it is used or who makes it.

The standard symbol for the crystal diode (such as geranium or
silicon) was adopted fairly recently and is now a part of the military
standard (Mil-Std 15A) governing all military publications. The actual
wording used in spelling out the accepted crystal diode symbol is as
follows: "The arrowhead designates the direction of forward (easy) current, as shown by a d-c ammeter. The arrowhead in this case shall be
filled." [1]

It is to be noted that the word "plus" is not used or needed in
identifying the diode elements by this standard method. While this
method is generally practiced in marking crystal diodes, the system has
not been generally followed in marking metallic rectifiers.[2]

Using type 1N34A as an example of a general-purpose diode, typical
results that may be expected are as follows:

Forward Resistance: around 400 to 500 ohms, as illustrated by for-
ward-current connection in Fig. 12-36.

Reverse resistance: generally in order of at least 100 times the forward resistance.

12-19. Transistor Checking by Ohmmeter

For checking transistors, a resistance check with an ohmmeter might
be used to check the forward and reverse resistance of its three elements,
taking two elements at a time. In such a case, the transistor is being
treated as if it were made up of two diodes; one diode considered as
made up of a base-to-emitter pair, and the other diode considered as
an emitter-to-collector pair. Such a method, though capable of detecting
opens or shorts, is much over-simplified, in that it disregards the transistor action as a triode. It is possible to check the triode action of transistor with an ohmmeter more effectively by means of a special method,[3]
the highlights of which are given below.

The operation of the transistor triode check is no more complicated
than taking two ohmmeter readings, but the special value of this method
lies in the fact that the transistor is energized by the battery of the

---

[1] Military Standard 15A, Washington, April 1, 1954.

[2] For a discussion of those rectifiers which still bear the "plus" marking on the component, see S. D. Prensky, "Much Fuss About Plus," *Radio-Electronics Magazine,*
April, 1956.

[3] S. D. Prensky, "Multimeter Transistor Checker," *Radio-Electronics Magazine,*
August, 1956.

ohmmeter during the process of taking two resistance readings, and that, as a result, the comparison of these two readings can be interpreted as a comparison of two transistor currents, thus yielding a measure of the current gain of the operating transistor..

Transistor Check Circuit. The test unit operates with the transistor in a grounded-emitter hookup shown in Fig. 12-37. This circuitry pro-

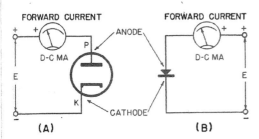

Fig. 12-36. Connection for forward-resistance check of: (A) tube diode, and (B) crystal diode.

vides one of the most significant clues to the condition of a transistor—i.e., its ability to amplify small changes in its base current. To do this, the test takes advantage of the circuitry in conventional service-type multimeters on their resistance ranges (shown in dashed lines). This section already contains a 1.5-v cell for voltage supply, a sensitive meter and a series-limiting resistor R_s, which protects both the meter and the transistor under test. Using these, the test circuit boils down to adding a few components externally, to provide a means for indicating changes in current, as the transistor is checked.

The first meter reading is taken with the spdt toggle switch in its LO position, corresponding to zero base-current bias. The toggle switch is then thrown to HI. This sends a small d-c bias current through the base of the transistor, causing a much larger current to flow in the collector circuit. The increase in meter reading indicates the relative current-amplifying ability of the transistor under d-c conditions. The difference between the high and low readings is then compared with the meter-swing increase caused by a good transistor of the same type. This provides a check on the condition of the transistor.

Transistor-Check Procedures With Various Ohmmeters. Since the transistor, like other semiconductors, is a nonlinear device, its effective resistance may vary widely, depending on the voltage across it. Thus, different ohmmeters, or more important, different ranges of the same ohmmeter, will result in noncomparable readings. This need not present any serious difficulties as long as only one ohmmeter and one resistance range of that meter is used; but it is well to understand the limitations. The meter used in this example is a Precision model 120. On its $R \times 100$ resistance range it uses a 1.5-v cell and has a center scale of 2,000 ohms.

Any other ohmmeter, using not more than 4.5 v and having a center scale reading of between 1,000 and 3,000 ohms is also suitable, with the proper allowance made for expected readings as compared with those of a good transistor.

The polarities shown in the circuit diagram are for p-n-p transistors. *The positive polarity does not necessarily coincide with the red lead of any particular ohmmeter.* In the Precision model 120, as in many others, the red lead is connected to the negative terminal of the ohm-meter battery. It is a simple matter to check this polarity on a voltmeter. It can also be checked on a diode, known to be good, by observing which polarity gives the conventional easy current flow for forward resistance.

For the p-n-p transistor check, a negative polarity is applied to the collector. With the switch at LO there is zero base current and the meter indicates the so-called "saturation current" (I_{co}) flowing in the collector circuit. When the switch is thrown to HI, the base is connected to a sufficiently negative point on the voltage divider (R_1 and R_2) so that

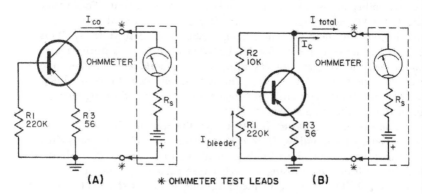

NOTE: The meter resistor R_s and battery are
integral parts of the ohmmeter, used on its RX100 range

Fig. 12-37. Test circuit for checking transistors with ohmmeter: (A) in LO position; (B) in HI position.

a small bias current (about 50 μa) flows through the base circuit. This results in a greatly increased collector current for the second reading. The difference in the two readings indicates the approximate relative effectiveness of the transistor for current amplification.

If an n-p-n transistor is to be tested, reverse the ohmmeter leads. The simple flipover automatically insures that the bias and collector circuits are connected to their proper polarities for the n-p-n transistor. The meter will still read in the forward direction and readings will be equally valid for this type of transistor.

Transistor Check Results. Table 12-1 gives a summary of results obtained with good transistors (and two defective ones, as noted), using the test unit connected to a Precision model 120 *VOM*.

After the first column of the table, identifying the transistor group, typical meter readings are given in the next two columns for the LO and HI positions of the test switch. The difference between these two gives the *net change,* a measure *of relative sensitivity or current amplification.*

If some other meter is used, the results of the table can easily be adapted, provided the meter's $R \times 100$ resistance scale is reasonably sensitive. The internal battery of the meter on this range should be between 1.5 and 4.5 and the center scale for the same $R \times 100$ range should read between 1,000 and 3,000 ohms.

TABLE 12-1. TRANSISTOR CHECK RESULTS*

| *Transistor group* | *Average meter readings*** | | | |
	LO	HI	*Net change*† HI - LO	*Range of net change*
GOOD TRANSISTORS:				
General purpose	6	40	34	24–40
R-f	3	38	35	34–37
A-f	5	43	38	37–39
BAD TRANSISTORS (GENERAL PURPOSE):				
Sample 1‡	6	18	12	
Sample 2§	1	7	6	

* From "Multimeter Transistor Checker," by S. D. Prensky, *Radio-Electronics Magazine,* August, 1956.

† For a workable minimum, a transistor should give a net change of at least 50% of the net-change figure to be considered good.

‡ Had been exposed to excessive soldering heat. Still showed satisfactory back-to-forward resistance.

§ Excessively high resistance in both forward and reverse directions.

** Based on a scale of 60.

13

APPLICATIONS IN RADIO TRANSMITTERS

13-1. Oscillators

A diagram of a typical electron-coupled oscillator used in an a-m transmitter is shown in Fig. 13-1 (A). Normal measurements are grid current with a low-range d-c milliammeter, and screen and plate currents with ranges suited to the tube and operating conditions. Meter 1 is frequently a permanent part of the transmitter circuit, since it is used in tuning the plate circuit to resonance. When the adjustment of capacitor C brings the tuned circuit near resonance, the plate current dips and becomes minimum at the resonant point. Voltage measurements are shown by the dotted lines.

Figure 13-1 (B) shows the circuit of a crystal oscillator. Similar measurements are made here, sometimes with the addition of an r-f meter connected in series with the crystal to check the r-f current through it. Since excessive r-f current through the crystal can cause damage, measuring keeps a check on it. Voltage measurements are shown by the dotted lines.

13-2. R-F Amplifiers in Transmitters

Figure 13-2 shows a class C amplifier used in radio transmitter r-f sections. As Oscillators do, class C amplifiers draw grid current. Therefore, meters must be used to measure the grid voltage or grid current. Note that a meter is shown in the *cathode* lead. The cathode current includes plate, screen, and grid currents, since they must all pass through the cathode in returning to the tube. The plate and screen voltages are often controlled by a switch; then, when the switch is off, the grid current alone registers on the cathode meter. However, it must

177

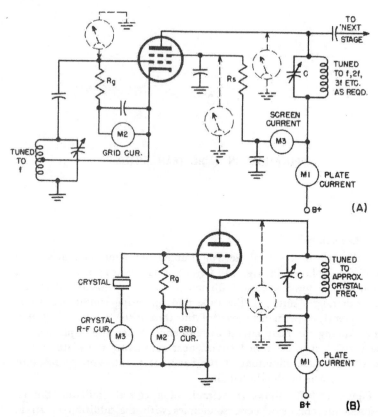

Fig. 13-1. Oscillator circuits: (A) electron coupled circuit such as is used for the fre-
quency-controlling circuit of transmitters, and locations of typical meter measurements;
(B) crystal oscillator circuit and meter-measuring locations.

be remembered that this grid current is somewhat higher than that
which actually flows when the plate and screen voltages are applied.

As in previous cases, screen and plate currents are measured by
series milliammeters connected as shown. Another meter indicates the
combined screen and plate current.

The circuit of Fig. 13-2 is general, and also applies to buffers,
frequency multipliers, and power amplifiers of the r-f sections of transmitters. The use of a triode, tetrode, or pentode determines the number
of active tube electrodes, hence the number of circuits which are
metered.

13-3. Using Meters in Neutralizing Transmitters

Most triode r-f amplifiers, and some tetrode and pentode amplifiers,
operated at very-high radio frequencies must be *neutralized*. This neces-

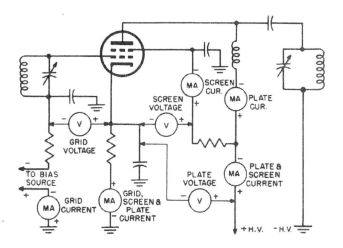

Fig. 13-2. Class-C r-f amplifier for transmitter, current, and voltage measurements.

sity arises from the fact that, without neutralization, the grid-plate interelectrode capacitance of the tube forms a path for positive feedback which causes the tube to oscillate, destroying its amplifier action. A neutralizing capacitor forms another path which feeds back energy in opposing phase to compensate.

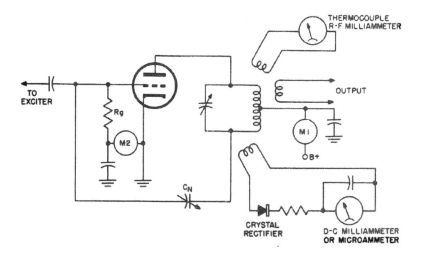

Fig. 13-3. Two methods of using meters to indicate r-f in the plane circuit of a class-C amplifier during neutralizing procedure.

Because of the variations in individual circuits, the neutralizing capacitance cannot be preset to the exact value necessary, but must be adjusted after the transmitter is constructed and being tested. Figure 13-3 shows a typical neutralized amplifier and two ways in which meters can be used as neutralization indicators.

The plate voltage is removed from the amplifier, but the exciter is left running, so that drive voltage is applied and grid current flows. If the circuit is not completely neutralized, some of the grid-circuit signal feeds through the grid-plate capacitance to the plate tank circuit. As C_N is tuned to the proper value, compensating grid signal is also fed through it to the plate tank. Thus, the capacitor C_N is adjusted for minimum r-f current in the plate tank. To make this adjustment possible, some indication of the r-f current in the plate tank must be available.

Fig. 13-4. Simple type of r-f wattmeter.

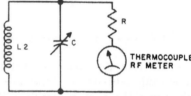

One type of indicator is a loop of wire whose ends are connected to a thermocouple milliammeter as shown. The r-f energy is coupled into the wire loop when the latter is held close to the plate tank coil and registers on the meter.

Another indicator uses a d-c meter movement in series with a crystal-diode rectifier. The crystal rectifies the r-f and the rectified current actuates the d-c meter. In the case of either indicator, the capacitor C_N is adjusted for minimum meter reading, while the plate tank capacitor is readjusted for resonance each time.

13-4. Measuring Transmitter Power Output

If the impedance of the antenna to which a transmitter is connected is accurately known, the r-f current into the antenna can be measured with a thermocouple ammeter and the square of the current multiplied by the antenna resistance to give the output power.

However, antenna impedance often is not known accurately, and transmitter capabilities are usually measured by loading the output stage with a *dummy load,* which is a load resistor designed to substitute for the desired impedance of the antenna. Devices used to measure transmitter power ouput are sometimes referred to as *r-f wattmeters.*

Perhaps the simplest r-f wattmeter is that illustrated in the diagram of Fig. 13-4. R-f energy is coupled from the tank circuit of the transmitter through *L1* into *L2.* The variable capacitor *C* is tuned to resonance

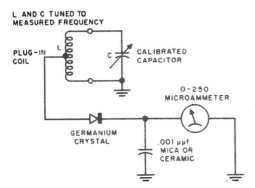

Fig. 13-5. Wavemeter using a microammeter and crystal diode to indicate resonance.

for maximum deflection of the thermocouple meter pointer. The power output then equals the square of the indicated r-f current times the resistance R. *R must be a non-inductive resistor whose r-f resistance at the frequency of measurement is accurately known.* It is also desirable that the value of R approximate the resistance of the antenna to be used.

13-5. Frequency Measurement with Wavemeter

Figure 13-5 shows how a d-c meter may be used with a crystal diode in a wavemeter to indicate resonance in making frequency measurements. The coil is coupled to the transmitter tank and the calibrated capacitor is adjusted until the meter reads a maximum. The calibration on the capacitor dial then indicates the frequency of the measured signal. Wavemeters are usually of the plug-in type and are equipped with a number of coils which can be changed to cover a wide frequency range.

13-6. Modulation Percentage

Figure 13-6 shows an instrument suitable for measuring modulation percentage. L and $C3$ are tuned to the transmitter carrier frequency.

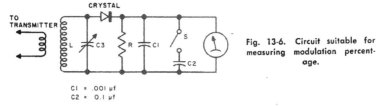

Fig. 13-6. Circuit suitable for measuring modulation percentage.

The crystal diode rectifies the r-f current so that a d-c voltage plus a-c modulation appears across R. $C1$ is a relatively small-value capacitor and filters out the rf. The d-c meter then responds to the average voltage. When S is closed the meter becomes peak-reading, and the d-c voltage indicated goes up to the peak value of modulation. The difference between the two readings compared to the d-c voltage (average value) is the percentage of modulation.

13-7. Grid-Dip Meter

One of the most useful instruments for design and testing of transmitters and receivers is the grid-dip meter. The circuit is shown in Fig. 13-7. It is simply a low power oscillator in which the grid current is measured by a sensitive microammeter. When the oscillator is loaded, such as by a nearby tuned circuit on the same frequency, the grid current is reduced. Thus, as the variable capacitor is adjusted so that the frequency of the oscillator passes through the resonant frequency of the

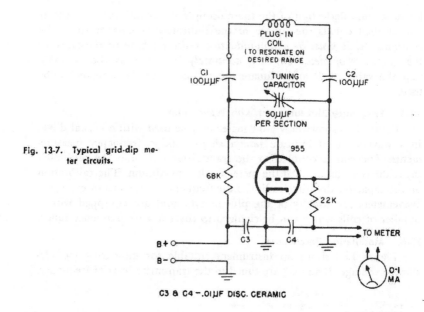

Fig. 13-7. Typical grid-dip meter circuits.

circuit to which the coil is coupled, the grid current "dips." The dial connected to the variable capacitor is calibrated to read frequency. Thus, the resonant frequency of a circuit in a transmitter or receiver can be determined.

The applications of the grid-dip meter are numerous. Given the coils which afford the required frequency range, the device is a low powered test signal source where such signals are required; when a system is tuned to its frequency and coupling exists between that circuit and the grid-dip meter, the meter indication will decrease abruptly at resonance.

14

SPECIALIZED MEASUREMENTS

14-1. Measurement Principles in Specialized Test Instruments

There are numerous instances where basic measurements are modified for special purposes, and these form a substantially large group of specialized electronic test instruments. While the details of such instruments are more properly discussed in a book dealing specifically with electronic testing instruments, it will be profitable in this chapter to investigate some fundamental measurement methods that are common to this large group of meter-using speciality instruments. The topics selected for this chapter are those whose principles of measurement cut across the three major divisions into which the over-all electronic field is generally considered to be divided—i.e., the power field, the communication field, and the industrial electronic-control field. These types of special measurements will be discussed in separate sections as follows:

a. Electronic-control measurements

b. High-sensitivity a-c measurements (employing both vacuum-tube and transistor amplifiers)

c. High-sensitivity d-c measurements (including electro-meter and micro-microammeter types)

d. Digital-counting type (including pulse counters and digital display measurements)

14-2. Electronic-Control Measurements

The field of electronic control (also generally termed "Industrial Electronics") makes wide use of the principle that mechanical quantities may be converted into electrical values, which are then usable in the form of electrical signals that can be easily measured and also utilized

183

for control purposes. The process of converting a mechanical quantity
into such an electrical signal is performed by a *transducer*, in the same
sense that a microphone acts as a transducer in converting sound-
pressure variations into an electrical signal varying at audio frequencies. Table 14-1 shows some of the outstanding examples in the
electronic-control field, where suitable transducers are widely used to
convert the nonelectrical quantity (shown at the left) into the electrical
instrument (at the right); the quantity of interest is usually displayed
on the meter of this instrument.

Although it is not feasible in a book primarily concerned with
meters to pursue each of these applications in detail, it is important
to realize that practically all of the high-sounding instrument names
given in Table 14-1 represent, in actuality, a basically similar process
of measuring an electrical quantity (usually voltage or its equivalent
in current) on a simple meter of one of the same types as have been
previously discussed. Even where the details become more complex
due to the addition of electronic circuitry to the basic meter, it will
be found that, for the most part, the fundamental principles of meter-
reading interpretation still apply.

In special cases where the transducer output voltage or current is
extremely small, as in the instances of thermocouples used for thermo-
meters and of phototubes used for light meters, the measurement
problem introduces greater difficulty than is present in the familiar
example of the microphone for sound measurement. Specific examples
are covered in later sections on specialized measurements in this chapter,
as in the case of very high-sensitivity a-c measurements and of extremely
high-impedance d-c measurements, such as are done by an electrometer.

14-3. High-sensitivity A-C Measurements

This type of instrument depends for its high sensitivity on the
ability of the electronic circuit to provide high-gain, stable a-c amplification without the drift problem attendant on d-c amplification. The
amplifier will, in general, require three or four pentode stages with
heavy feedback to achieve the required combination of stable-gain–
bandwidth product. The voltage indication in some instruments results
in a uniform scale of rms volts, and in others is displayed as a uniformly spaced scale, thus requiring a logarithmic response from the
indicating meter. A brief description of a commercial version of each
type is given in Sec. 14-4.

14-4. Rms-Reading Multistage Instruments

The circuit of this multistage a-c amplified type voltmeter includes
a two-step voltage divider in the input, a stabilized broad-band amplifier,

a rectifier and meter circuit, and a regulated power supply. Arrangement of the circuit is shown in block diagram form in Fig. 14-1, and in the partial schematic of Fig. 14-2.

The voltage under measurement is applied to the voltmeter at the input terminals. On the lower ranges the voltage is coupled directly to the grid of a cathode follower, the cathode of which is connected as

TABLE 14-1. ELECTRONIC CONTROL INSTRUMENTS

Nonelectrical quantity	Electrical measuring instrument
Sound pressure	Audiometer
Air or liquid pressure	Pressure Gauge
Temperature	Thermometer
Moisture content	Hygrometer
Acidity	pH-meter
Mechanical strain (displacement)	Straingauge
Velocity	Speedometer or tachometer
Acceleration	Accelerometer
Fuel level	Fuel gauge
Brightness	Light meter
Simulated mechanical variation	Analog computer (electrical variation)
Numerical quantity (as a pulse)	Digital computer (electrical pulse counter)

a voltage divider tapped for six outputs. On the higher ranges the voltage is reduced to a thousandth of its value at INPUT terminals before it is applied to the grid of the cathode follower. Out of the six-tap divider, the voltage is coupled to a four-stage amplifier. The output of the amplifier feeds into a full-wave rectifier bridge with a d-c milliammeter across its midpoint, and a current proportional to the input voltage flows through the meter movement. The meter is so calibrated that the resulting deflection indicates the rms value of a sine-wave voltage applied at the input terminals.

The method used for obtaining the rectified current to actuate the d-c meter is worthy of closer attention. For purposes of explanation, the amplifier (tubes V_2 through V_5) may be considered as a source of a-c signal, with the output from the plate of V_5 as one side and the cathode of V_2 as the other side.

The rectifier-meter circuit is arranged in a bridge-type configuration, with a crystal diode and a capacitor in each branch and a d-c milliammeter connected across its midpoints. The diode connection provides full-wave rectification of the input current. The design of the bridge is such that (1) a pulsating direct current is delivered to the meter

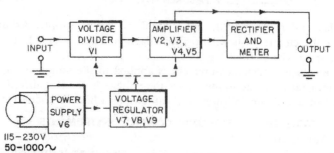

Fig. 14-1. Block diagram of a-c amplifier type VTVM. *Hewlett-Packard model 400D*

circuit, and (2) an alternating current of the same frequency as the
current at the rectifier input is delivered to the output of the bridge.
From the rectifier-bridge output, the a-c component flows through the
feedback loop to the cathode of V_2.

The current through the meter is proportional to the average value
of the waveform of the voltage applied to the input of the rectifier.
Since calibration of the meter in rms volts is based on the ratio that
exists between the average and effective values of a voltage that is a
true sine wave, deviation in a waveform from that of a true sine wave

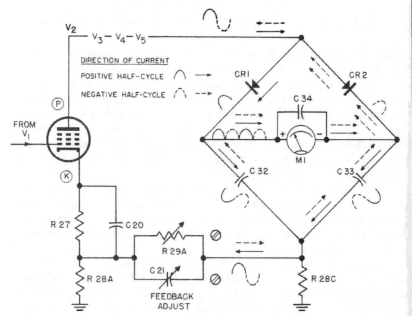

Fig. 14-2. Partial schematic showing indicating circuit in VTVM. *Hewlett-Packard
model 400D*

may cause meter measurements to be in error. Table 14-2 gives an indication of the limits of possible error due to the presence of harmonics in the waveform of a voltage under measurement.

TABLE 14-2. EFFECT OF HARMONICS ON MODEL 400D/H VOLTAGE MEASUREMENTS*

Input voltage characteristics	True rms value	Value indicated by model 400D
Fundamental = 100	100	100
Fundamental +10% 2d harmonic	100.5	100
Fundamental +20% 2d harmonic	102	100–102
Fundamental +50% 2d harmonic	112	100–110
Fundamental +10% 3d harmonic	100.5	96–104
Fundamental +20% 3d harmonic	102	94–108
Fundamental +50% 3d harmonic	112	90–116

* Courtesy: Hewlett-Packard Company.

NOTE: This table is universal in its application since these errors are inherent in all voltage measuring equipment of the average-reading or peak-reading type.

14-5. Logarithmic-Reading Instruments

The logarithmic-scale voltmeter employs a resistance-capacitance decade attenuator, a high-gain, three-stage, negative-feedback, stabilized amplifier, a rectifier, and a logarithmic indicating meter responding to the average values of the voltage wave, but calibrated in rms values of a sine wave. The amplifier has sufficient feedback to render the indications of the voltmeter substantially independent of changes in line voltage, tubes, and other circuit components.

By means of a five-decade range switch, the entire voltage range of 1 mv to 100 v ac is indicated on a single logarithmic voltage scale reading from 1 to 10. The indicating meter also has a uniform db scale, numbered from 0 to 20 db. Its accuracy is 2% from 10 to 150,000 cycles. Since the scale is logarithmic, the accuracy is the same at any point on the scale.

In a meter of the logarithmic-reading type, as shown in Fig. 14-3, there is no zero-volt marking on the scale, and therefore, as was previously mentioned for meters of this type, it is to be expected that the pointer will rest off-scale at the left when the voltmeter is in an unenergized condition.

14-6. Transistorized Voltmeters

The development of voltmeters employing dependable transistor amplifiers fulfills a long-felt need for a sensitive meter that is at once free of a-c hum pickup (because of its eminently practical battery operation) and, at the same time, is capable of extremely high amplification. When compared to a vacuum-tube voltmeter (VTVM) in these respects, the transistor voltmeter (TRVM) exhibits superiority in that the transistors do not require heater current, and their collector-current drain is so small that battery operation is entirely feasible and in fact

Fig. 14-3. Logarithmic-type meter. *Ballantine a-c VTVM model 300*

often preferable to line operation even when a-c power is readily available. By the use of the transistor amplifier, voltage sensitivity down to 1 mv (1×10^{-3} v) full-scale is provided in the first example cited below, and with special precautions, a full-scale sensitivity of 10 μv (1×10^{-5} v) is obtained by the second example cited. Such measurement sensitivity is not a standard capability in vacuum-tube voltmeters.

The theory and practice of transistor-amplifier operation is beyond the scope of this book; detailed treatments will be found in the reference books listed at the end of this chapter. We are concerned here only with one important limitation in transistor operation; i.e., the comparatively low input-impedance of a transistor amplifier, as compared with that of a vacuum tube. This limitation is inherent in the fact that the transistor is a current-actuated device in contrast to the vacuum-tube which is essentially voltage-operated. As a result, input impedances beyond 1 megohm are not readily attainable with transistors, even in the common-collector configuration, which generally provides input impedances of the order of tenths of a megohm. One method for overcoming this limitation makes use of a vacuum tube operated as a cathode follower, resulting in a hybrid (tube and transistor) instrument. This method is discussed in the following example of a commercial instrument (see Fig. 14-4).

SPECIFICATIONS FOR THE ALTO MODEL D-21 TRANSISTORIZED A-C VOLTMETER

Voltage ranges, full-scale: 0.001 v (1 mv) to 300 v in 12 ranges.

Frequency range: 10 cps to 1 mc.

Operating temperature range: 0°F to 120°F.

Accuracy: within ±5% of full scale.

Input impedances: 10 meg and 15 $\mu\mu$f, 1 to 300 v; 10 meg and 30 $\mu\mu$f, 0.001 to 0.3 v.

Output: undistorted amplifier output voltage is 0.05 v (open circuit) corresponding to 36 db max.

Transistor, tube, and battery complement: one subminiature pentode, CK6611; four transistors, p-n-p surface-barrier, SB100; four mercury batteries, 9.4 v each; one mercury cell, 1.34 v.

Battery life: approximately 130 hr continuous duty.

Fig. 14-4.   Transistorized a-c voltmeter. *Alto model D-21A*

The voltmeter circuitry illustrated in the block diagram (Fig. 14-5) is seen to consist of the tube cathode follower which is both preceded and succeeded by the range attenuator and which feeds the four-transistor a-c amplifier. The amplifier output is rectified by a bridge rectifier which actuates the indicating meter and also provides inverse feedback to stabilize the sensitive amplifier circuit. Further stabilization and transistor interchangeability is accomplished by negative feedback in each stage, introduced by the use of unbypassed emitter resistors.

The partial schematic diagram of Fig. 14-5 shows the final transistor stage and output arrangement. The a-c output is taken from the collector of transistor $Q4$, operated in a common-emitter configuration. However, the inclusion of the unbypassed emitter resistor of 100 ohms allows the a-c amplifier output to be taken separately from the emitter as a low-impedance output (50 mv max). The full-wave rectifier bridge

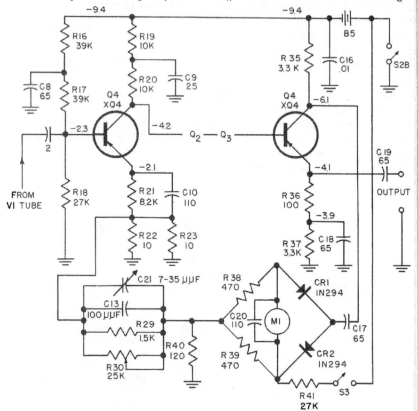

Fig. 14-5. Partial schematic of transistorized voltmeter circuit. *Alto model D121A*

uses two crystal diodes and two resistors in its arms, and feeds the 200 μa d-c indicating meter. Part of the output of this last stage is fed back to the emitter of the first stage, providing feedback over all the stages. The total amount of feedback is approximately 38 db at mid-band, which ensures high stability of the voltmeter over the wide temperature and frequency range.

Of the five batteries used to supply power, one is used to supply 1.3 v for the filament of the tube, three 9.4-v batteries in series supply

28.2 v for the tube plate, and the final 9.4-v battery supplies the required voltage for the transistor amplifier. A push-button test is incorporated to show battery condition and to indicate when battery replacement becomes necessary. The operating performance of the mercury battery is well suited to test-instrument use, since full advantage can be taken of the superior constant-voltage and long shelf-life characteristics of mercury cells.

It is interesting to consider another example of a commercial transistorized voltmeter (Millivac model MV-45A), which tackles the noise problem in ultrasensitive voltmeters by the use of a special "hushed-transistor" principle, i.e., operation with a drastically reduced noise level 20 db or more below the noise level of the best vacuum tubes. This makes it possible to accomplish a full-scale range as low as 10 μv in this all-transistor voltmeter, which employs a total of nine transistors. However, the extreme sensitivity obtained by all-transistor operation entails a corresponding compromise with input impedance. This is approximately 10 k on its 10-μv range, rising to 1 meg in the millivolt range. Since no tubes are employed, the current drain from its 6-v battery supply is only 12 ma, resulting in a low power consumption of only 72 milliwatts and hundreds of hours of battery life.

14-7. High-sensitivity D-C Measurements

The term "high sensitivity" as applied to d-c measurements, embraces those instruments having a sensitivity substantially greater than the 10 megohms-per-volt value that is generally associated with the conventional service-type VTVM. Table 14-3 lists the comparative sensitivities of the major d-c voltmeter types in approximate values.

It may be seen from the above comparison that, between the least and the most sensitive types listed, there is a very great range of difference, the highest ohms-per-volt value of the electrometer type being around (10^9) or 1 billion times greater than that of the 1,000 ohms-per-volt value of the moving-coil meter prototype, with the 10 megohm-per-volt value of the service-type VTVM (of the type discussed in Chap. 9) falling about midway between the two extremes. Since the circuit loading, i.e., the current taken by the instrument from the source being measured, for a given full-scale reading, is directly related to the ohm-per-volt value, it can be seen that measurements with the high-sensitivity instruments can be made, having many thousands of times less disturbing effect on the circuit, than would be caused by the use of an ordinary VTVM. In spite of their higher cost, therefore, these high-sensitivity instruments find application in special circuits where *an extremely high impedance,* such as insulation resistance, is being measured or in other cases where the measurement of *very minute currents,* such as grid currents and photo currents, is required.

The electrometer type and the d-c modulated type of high-sensitivity d-c measurements will be briefly considered in separate paragraphs that follow.

14-8. Electrometer Instruments

Electrometer tubes are operated at low plate voltages and are specially designed to keep the grid current down to fractions of micro-microamperes; 1 $\mu\mu$a (1×10^{-12} amp) is a typical value of the grid current, which is about 1/10,000 the grid current present in the ordinary vacuum tube. As a consequence, very high values of grid resistor may be used, providing an input resistance that may exceed millions of megohms (megamegohms or 10^{12} ohms). Figure 14-6 is a commercial version and its simplified input circuit is shown in Fig. 14-7. In the d-c VTVM, the electrometer tube is combined with a multistage direct-coupled amplifier, which is stabilized by the use of a large amount of degenerative feedback and a highly regulated power supply. The resulting combination provides a very high oms-per-volt sensitivity, in the order of a million megohms-per-volt, for the measurement of d-c voltages down in the millivolt ranges. By way of illustration, the Keithly d-c VTVM (model 220) measures d-c volt full-scale

TABLE 14-3. COMPARATIVE D-C SENSITIVITIES

Instrument	Resistance, ohms/volt	Full-scale range, volts	Input impedance, ohms	Current taken by instrument, amps
VOM:				
prototype	1 k	1	1 k	1 ma
	(1×10^3)		(1×10^3)	(1×10^{-3})
service-type	20 k	1	20 k	50 μa
	(2×10^4)		(2×10^4)	(5×10^{-5})
VTVM:				
service-type	10 meg	1	10 meg	10 μa
	(1×10^7)		(1×10^7)	(1×10^{-7})
HIGH-SENSITIVITY:				
electrometer	1,000,000 meg	0.1	100,000 meg	1 $\mu\mu$a
	(1×10^{12})	(1×10^{-1})	(1×10^{11})	(1×10^{-12})
d-c modulated	100,000 meg	100 μv	10 meg	10 $\mu\mu$a
	(1×10^{-11})	(1×10^{-4})	(1×10^7)	(1×10^{11})

from 30 mv to 100 v, at an input resistance of 100 megamegohms (1×10^{14} ohms). With appropriate accessories, voltages down to 1 mv and correspondingly very small currents (5×10^{-14} amp) can be measured, while resistances as high as 10,000 megamegohms (1×10^{16} ohms) can be adequately handled.

14-9. D-C Modulated Type

The difficulty with drift that is inherent in the direct-coupled amplifier of the electrometer type, is overcome to a large degree in the d-c modulated system by the use of a chopper-vibrator or other modulating scheme to convert the d-c input to an a-c carrier signal which is amplified by a stabilized multistage a-c amplifier and then reconverted to dc to activate the indicating meter.

Because of the inherent drift limitations, which is in the order of millivolts, the smallest full-scale range of the d-c amplifier, (using round numbers), is limited to an input signal of around 0.1 v. In the chopper-modulator system, the drift can be cut down by a factor close to 1,000, thus allowing an input-signal range of around 0.1 mv or 100 μv) full-scale to be handled. This advantage of sensitivity to smaller voltages, however, is obtained at the expense of a lowering of input impedance necessitated by the dc to ac conversion. As a consequence, the over-all ohms-per-volt sensitivity is not substantially altered, as may be seen from the comparison chart where both types wind up with ohms-per-volt ratings in the order of megamegohms per volt (from 10^{11} ohms to 10^{12} ohms per volt). The important distinction that may be drawn from these figures is that, while the electrometer type of instrument provides the highest input impedance for d-c measurement, the d-c modulated type provides the lowest full-scale range of measurement, both meters having similar ohms-per-volt capabilities.

Fig. 14-6. Electrometer. *Keithley model 210*

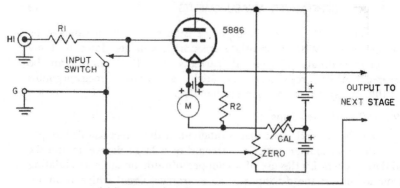

Fig. 14-7. Simplified input circuit of electrometer. *Keithley*

A commercial version of the chopper–ac-amplifier type of d-c modulated instrument, is illustrated in Fig. 14-8. This model has a zero-center scale, making it useful both as an electronic galvanometer and as a d-c microvoltmeter.

Typical specifications for the chopper-modulated type of d-c instrument are as follows:

SPECIFICATIONS FOR THE KINTEL D-C MICROVOLTMETER MODEL 203

Voltage range, full-scale: 100 μv to 100 v.

Current range, full-scale: 100 $\mu\mu$a to 100 ma.

Input impedance: Below 10 mv, 10 meg; at 30 mv, 30 meg ;above 30 mv, 100 meg.

Rating as amplifier: 80 db gain; 1 v output across 100 ohms.

Drift, after 15 min warm-up: 10 μv equivalent input.

Output impedance: less than 2 ohms

The input, when applied to the chopper amplifier in the form of a d-c or slowly varying signal, produces a proportional a-c carrier signal. This a-c signal is amplified by a highly stable, inverse feedback multistage amplifier and is then synchronously rectified to produce a large proportional d-c signal output.

14-10. Digital-Display Measurements

The digital-display type of instrument does not possess the usual scale and pointer on its face. Instead, it displays the measured quantity (such as voltage or resistance) in terms of a series of numerals separated by a decimal point. Thus in a representative three-digit type of VTVM (Fig. 14-9) a low voltage might be displayed by a number such as 0.01, up to a high-voltage limit of 999 v. The main advantage in having the value displayed on a counter is not just the ease of reading, but more important, it provides an output in a form inherently suitable for read-out from a computer and also suitable as a subsequent input to a succeeding section of the computer. In this respect, it should be considered a specialized instrument whose extra cost is unquestionably warranted in computer use or other special application, but is ordinarily not justified for general-purpose use.

The consideration of pulse-counting circuits in general is a broad topic which includes counter instruments having widely different uses. They all display a numerical value based on a pulse count proportional to the measured quantity. The quantity being measured may be, for

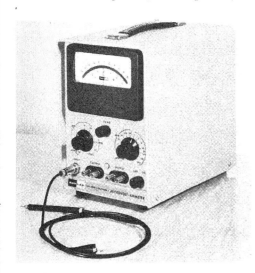

Fig. 14-8. D-c chopper-modulated microvoltmeter. Key Lab (Kintel) model 203

example, frequency in the case of frequency-meters; or in the case of ionization events produced by radioactivity, it may be counts per minute or milliroentgens per hour, this constituting an instrument for measuring radioactivity. These specialized instruments would more properly be discussed in special texts on the subject, such as some of the references given at the end of this chapter. Nevertheless, because of their increasing use in electronic data processing (EDP) computers, it is well to examine the general principles upon which these instruments operate.

The digital types of meter producing a numerical display of some electrical quantity, may be roughly divided into two types: electromechanical and all-electronic. This feature will be considered here from the standpoint of its special utility in measurements requiring such a form of read-out as, for example, in the field of computers. A further distinction will be made between two electromechanical types that differ in the source of the impulses that drive the digital counter unit. These impulses may be obtained from either a servomotor or a stepping switch. Both the electromechanical and the all-electronic models operate on the same basic principle of voltage comparison.

In general, the digital voltmeter operates as a self-balancing nulling device to produce the digital display of the voltage being measured. It is based upon the basic "potentiometer" principle—here using the

Fig. 14-9. Digital VTVM. Hycon model 615

term "potentiometer" in its historical sense of a potential-nulling device and not in the more popular sense of voltage divider. To clarify this discussion, the term "potential-nulling device" will be used when the historic type of "potentiometer" is meant, and the term "voltage-divider" will be used to indicate the ordinary three-terminal type of variable resistor in which the moving arm picks off a portion of the full voltage applied to its outer terminals.

The circuit of the basic potential-nulling device is shown in Fig. 14-10 where an unknown voltage E_x is opposed by a portion of the reference voltage selected by the arm of the voltage divider E_{rp}. When the selected portion E_{rp} is equal and opposite to the value of the unknown voltage E_x a null-indicating device, (such as a galvonometer) will indicate zero, and the desired voltage value can then be obtained from the calibration on the voltage-divider dial. This method of reading voltage, even though an indirect one, possesses the important advantage of drawing zero current from the unknown voltage, when the reading is taken at the exact balance condition. (This is the same principle as that used in the slide-back type of voltmeter.) At that time, the unknown (E_x) can be computed from reference voltage (E_{VD}) with the formula:

$$E_X = \frac{R1}{R1 + R2} \; E_{VD}$$

When the unknown voltage (E_x) is greater than that from the voltage-divider (E_{VD}), the galvanometer deflects in one direction; and when it is smaller than E_{VD}, the pointer swings to the opposite direction. As a result, a change in the polarity of the error signal is produced, depending upon whether the unknown voltage is greater or smaller than the selected portion of the reference. It is the change in error signal which provides the driving force that finally produces the digital display of the unknown voltage.

In practice, the error signal is not obtained from a galvanometer, but is obtained instead from the output of a chopper or vibrator, which alternately contacts first the unknown-voltage side and then the

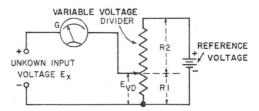

Fig. 14-10. Basic "potentiometer" (potential-nulling) circuit.

voltage-divider side, so that an error voltage is developed in proportion to the difference between the two sides. In effect then, the chopper serves as a voltage comparator whose difference-signal output is fed to an amplifier. Consequently, when the two voltages being compared are equal in value, the a-c output of the chopper-amplifier will be zero, and this output will become increasingly greater as the inequality of the two voltages becomes larger. The magnitude of this amplified error-signal acts as the driving force that produces the final display in either of the two electromechanical systems.

The form of the three-digit VTVM illustrated in Fig. 14-9 uses the action of a servo motor to actuate the digital counter unit. A simplified diagram of this system is shown in Fig. 14-11 for the d-c volts function.

With no voltage at the unknown terminals, the voltage-divider arm in Fig. 14-11 is normally in the position which places it at ground potential, with the servo motor at rest and its indicating dials at 000.

The input to the servo motor is the amplified error signal, which is thus zero at the start. The motor, it will be noted, is mechanically coupled to the moving arm of the voltage divider. When the unknown voltage is applied to the left-hand chopper input (generally through the range attenuator), the servo motor drives the counter and simultaneously drives the moving arm in such a direction as to reduce the

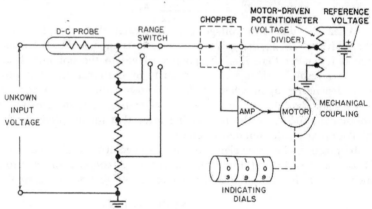

Fig. 14-11. Simplified diagram of servo-motor action in d-c circuit of digital VTVM. Hycon model 615

original error signal. When the arm (often called an answer arm) has moved to a position that reduces the error to zero, the amplifier output also falls to zero, and the servo motor stops. At this point the counter digits indicate the value of the unknown voltage.

The placement of the decimal point and the signal polarity are indicated by lamps which illuminate appropriate symbols on the front panel of the instrument. If the test voltage is too high for the scale in use, a limit switch keeps the dials from rotating more than 40 digits beyond 999.

If the polarity of the input signal does not agree with that indicated by the input selector switch, the servo motor tends to turn the indicating dials backwards. If this occurs, mechanical and electrical stops keep the dials from moving too far to prevent damage to the mechanism. If the polarity of the input signal agrees with that of the input selector, the motor rotates the indicating dials and the precision potentiometer in the proper direction. Since the input selector switch reverses the polarity of the mercury reference cell as well as the polarity of the reference-phase winding of the servo motor, the polarity at the potentiometer arm will always match that of the input signal, if the input selector is in the proper position.

Resistance measurements are made in a similar fashion by means of a Wheatstone-bridge circuit.

The other electromechanical type of digital VTVM is one in which the counter dials are actuated by stepping switches, rather than by a servo motor. The same basic principles of voltage comparison, chopper sampling, and error-voltage amplification are again used here, as in the previous example. The main difference lies simply in the fact that stepping switches are employed here instead of servo motors; but the mechanical coupling is still retained between the positioning ele-

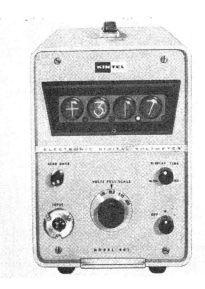

Fig. 14-12. All-electronic digital VTVM. *Kintel model 801*

ment (in this case the stepping switches) and the voltage-divider arm supplying the "feedback" or "answer" element. While the resulting display is in digital form in either case, the latter instrument, employing stepping switches, incorporates an additional provision for supplying read-out pulses for computer operation, simultaneous with the display of voltage.

The all-electronic type illustrated in Fig. 14-12 is also based on a trial-voltage comparison method and employs a beam-switching technique. It provides a digital in-line read-out of d-c voltages in 0.1 sec with high accuracy (0.1%); special models of this type provide binary-coded decimal and decimal outputs for driving typewriters or other recording devices, while simultaneously providing digital monitoring.

In the method of trial-voltage comparison (sometimes called a successive-approximation method), after a number of successive voltages from the trial-voltage generator have been compared with the input voltage, a logic network activates the glow-tube ("Nixie") display. For the three-digit unit (Hycon model 801) the comparison process itself is accomplished in approximately 60 msec. Its display time is adjustable from 0.1 sec to infinity.

REFERENCE LIST FOR SPECIALIZED MEASUREMENTS

The examples of commercial instruments given in this chapter amply indicate that the special nature of many electronic measurements requires further study in specialized texts. Here follows a list of some outstanding books published by John F. Rider on these special measurements, as well as a list of some special texts from other publishers on the subjects of electrical and electronic measurements.

Rider Books

A. Ghirardi and R. Middleton, "How to Use Test Probes," 1955.
Hazeltine Corp. Lab Staff, "Color Television Receiver Practices," 1955.
J. R. Johnson, "How to Use Signal and Sweep Generators," 1954.
A. Lytel, "UHF Practices and Principles," 1954.
D. Mark, "Basics of Phototubes and Photocells," 1956.
H. E. Marrows, "Transistor Engineering Reference Handbook," 1957.
J. S. Murphy, "Basics of Digital Computers," 1958.
S. Platt, "Industrial Control Circuits," 1958.
J. F. Rider, "The Vacuum-Tube Voltmeter," 1941.
J. F. Rider, "Obtaining and Interpreting Test Scope Traces," 1955.
J. F. Rider and S. D. Uslan, "FM Transmission and Reception," 1951.
Van Valkenburgh, Nooger, Neville, Inc., "Basic Synchros and Servo-mechanisms," 1955.
Zbar and Schildkraut, "Advanced Television Servicing Techniques," 1954.

Books on Electrical Measurements

Melville B. Stout, "Basic Electrical Measurements," Prentice-Hall, Englewood Cliffs, N. J., 1950.

Forest K. Harris, "Electrical Measurements, John Wiley, New York, 1952.

Walter C. Michels, "Electrical Measurements and their Applications," D. Van Nostrand, New York, 1957.

Ernest Frank, "Electrical Measurement Analysis," McGraw-Hill, New York, 1959.

Books on Electronic Measurements

Rufus P. Turner, "Basic Electronic Test Instruments," Rinehart Books, New York, 1953.

F. E. Terman and J. M. Pettit, "Electronic Measurements," McGraw-Hill, New York, 1952.

Gordon R. Partridge, "Principles of Electronic Instruments," Prentice-Hall, Englewood Cliffs, N. J., 1958.

15

ADVANCED METER FEATURES

There are a number of additional features that may be added to increase the convenience and versatility of the meter as a measuring instrument. This chapter will consider refinements which, though not absolutely essential for meter function, nevertheless facilitate operation.

15-1. Automatic-Ranging Meter

In the automatic-ranging type of meter, there is no switch on which the range scale is selected. Instead, once the function to be measured has been set (say d-c volts, for example), simply applying the meter leads to the circuit being measured causes the meter to cycle itself automatically to the proper range. Where overlapping ranges are present, the automatic action will also select the range which gives a reading on the upper rather than the lower part of the scale, in the interests of better accuracy. An example of an automatic-ranging meter is seen in Fig. 15-1. This meter, as a representative example, produces automatic ranging by comparing the voltage being measured with various reference voltages, and produces an error signal which causes a servo motor to rotate the range switch until it reaches the appropriate contact, at which point the voltage being measured is within the range of the reference voltage connected to that contact point, and the servo-motor action ceases.

This feature has particular value in situations where the meter is being used by nontechnical personnel—especially production testing. The automatic-ranging feature, in such instances, allows the measurement testing to be performed rapidly and on a nonskilled basis.

15-2. Suppressed-Zero Measurements

In these measurements, we are primarily interested in a restricted voltage variation, such as a line-voltage variation, for example, between only 105 and 125 v, rather than the full variation between zero and

maximum. It is an obvious advantage here to have a meter whose
pointer will indicate only between the limits of interest and will thus
allow the meter to be read much more easily and definitely. This is
accomplished by a suppressed-zero type of meter, which was introduced
previously in Chap. 2. A commercial version featuring the open scale
is seen in Fig. 15-2.

The expansion of the useful portion of the conventional scale is
accomplished by means of a sensitive bridge network. This network is
arranged to be in balance at only one value of input voltage corresponding to the center-scale value, which in this case is 115 v. Any
deviation from this value results in a bridge unbalance, which then
results in a corresponding left or right motion of a standard microammeter movement.

The resulting accuracy of the expanded-scale meter is expressed
in percentage of the center-scale value. This in the case of a voltmeter
with a center-scale value of 115 v and a range between 105 and 125 v,
accuracy is stated as within ±0.6% of the center scale value, which
comes out to be ±0.7 v in this case. This is a decided advantage over a

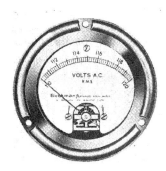

Fig. 15-2. Suppressed-zero type voltmeter. Beckman/Helipot model 026-3007-0

conventional meter, whose accuracy is stated in percentage of full-scale readings; in addition, more precise readings can be obtained with the suppressed-zero meter because of high resolution of the linearly expanded scale.

15-3.   "Burnout-Proof" Meters

An additional feature incorporated in some multimeters is shown in the "burnout-proof" type of meter illustrated in Fig. 15-3. This meter is protected on all of its ranges by a circuit-breaker action, which causes the red button shown on the right side of the meter to pop out and disconnect the circuit, whenever an overload occurs. This patented

Fig. 15-3. Burnout-proof multimeter. *Hickok model 455*

feature is so arranged that the circuit will be broken in time to give adequate protection even when extreme overloads occur, as in the case when the 117-v a-c line voltage might be applied directly across the meter's most sensitive μa range. The circuit action is also sensitive and fast enough to detect weaker overloads before the meter is damaged, and it does this on the ohms scale in addition to the voltage and current ranges. It can be seen from this that the circuit breaker used has a very wide range of operation. The circuit is restored to operation after the overload has been cleared by simply pressing the red button, thus restoring it and the circuit to their normal operating condition. This feature is particularly useful for beginners, and it may well be quite

valuable to the experienced operator also, in guarding against a careless
accident.

15-4. Pocket-Size Meters

A further contribution to the convenient use of meters is the introduction of the pocket-size meter type. The advantage in portability
of the meter shown in Fig. 15-4 is obvious, especially when it is considered that the size of the meter is only $3\frac{1}{2}$ by $5\frac{3}{8}$ by 1-11/16 in. In
spite of their small size, these meters have the same range of perform-
ance ability as their larger brothers, and are miniatures only in the sense
of having a smaller over-all length of scale. For example, the repre-

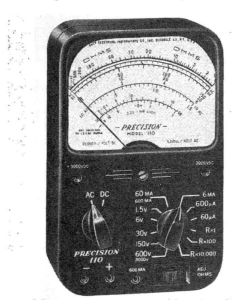

Fig. 15-4. Pocket-size multimeter. Precision Apparatus
model 110

sentative meter illustrated measures 5 a-c voltage ranges at 5,000 ohms
per v, 5 d-c voltage ranges at 20,000 ohms per volt and has 5 current
ranges from 60 μa to 600 ma full-scale. It incorporates three resistance
ranges from $R \times 1$ to $R \times 10,000$. The cost of these pocket-size types
is generally in the same range as the more conventional size meters,
since it turns out that any possible saving in the cost of materials is
balanced out by the added manufacturing difficulties caused by the
miniaturization. Other less expensive models using pinjacks rather than
a rotary switch for range selection are available. Both kinds of these
handy little meters have proven very popular.

INDEX

Lightning Source UK Ltd.
Milton Keynes UK
UKHW022032150223
417099UK00008B/115